Patrick Moore's Practical Astronomy Series

For further volumes:
http://www.springer.com/series/3192

Amateur Telescope Making in the Internet Age

Finding Parts, Getting Help, and More

Robert L. Clark

 Springer

Robert L. Clark
Westminster, MD 21157, USA

ISSN 1431-9756
ISBN 978-1-4419-6414-4   e-ISBN 978-1-4419-6415-1
DOI 10.1007/978-1-4419-6415-1
Springer New York Dordrecht Heidelberg London

Library of Congress Control Number: 2010935731

Printed on acid-free paper

Springer is part of Springer Science+Business Media (www.springer.com)

Acknowledgments

Many people should share in the credit or blame for this book, some for direct contributions and some for encouragement and support of the activities leading up to this book.

Of the first group my wife, Nancy, tolerated lots of things not getting done as time went to the book. She copied files and pictures, providing needed technical support. Gerry Frishkorn from the Westminster Astronomical Society provided the neat photo of a spider from his famous binoculars. He also checked and detected errors in my math. Numerous folks from the same club provided leads for the supplier appendix. Cousins Ann Herbert and Faith Marchal provided help with the conversion material. Cousin Hughes Gemmill gave me some old Dallmeyer lenses with which I learned how to get them unglued and cleaned.

In the second group are Duane Shie, who is inspiring as a dumpster diver; George Lockwood, who was always available for idea testing and eBay hints; Teresa Rosales; Blas Chavez; Roger Baker; and Mark Dennis, who helped with the lugging of telescopes and their parts.

The folks at Springer publishing have been very tolerant of the questions of a beginning author and have provided important guidelines.

Contents

About the Author

Although Clark was born in New Hampshire, he spent his early teen years in the New York City area. There he became acquainted with the Theodore Roosevelt Museum of Natural History and at the associated Hayden Planetarium, operated, at that time, by Newton and Margaret Mayhall. Their book *Skyshooting* provided his first introduction to the idea that someone with only moderate skills can build good scientific instruments. The three volumes of *Amateur Telescope Making*, edited by Albert G. Ingalls, completed the process of addiction.

Clark received a B.S. degree in Mathematics from Stanford University. He then served in the US. Air Force at Strategic Air Command headquarters in Nebraska. After his discharge he did graduate work in Mathematics and Computer Science at the University of Maryland. After a few years spent in the Military Operations Research industry, he migrated to a university teaching position that lasted 33 years at the University of the District of Columbia.

Throughout all this time he continued to be enamored by the process of building telescopes and actually built a few as time permitted. Since retirement he has allowed the addiction to occupy a large part of his enthusiasm, as is made evident by Fig. 2.3.

Clark now resides in Westminster, Maryland, where he has built a hilltop observatory and is active with the Westminster Astronomical Society.

Chapter 1

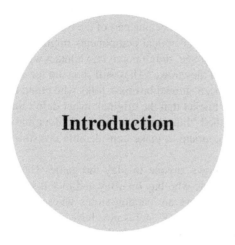

Introduction

Amateur telescoping making has been, since its beginning, an arena for inventiveness and fierce independence. Part of the credo of the amateur telescope maker (ATM) has been: "What I can't get (probably cheaply) from the everyday content of industry I will make for myself".

Another point that differentiates ATMs from average amateur astronomers is that they truly want to see the universe on their own terms, through an instrument of their own hands.

The following was said by the founder of Stellafane, Russell W. Porter, in March 1923. Stellafane is the name of the clubhouse built by the Springfield Telescope Makers club of Springfield, Vermont, in the early 1920s. It has since come to refer to the club's property and the convention, a yearly star party held each year at that location.

> For it is true that astronomy, from a popular standpoint, is handicapped by the inability of the average workman to own an expensive astronomical telescope. It is also true that if an amateur starts out to build a telescope just for fun he will find, before his labors are over, that he has become seriously interested in the wonderful mechanism of our universe. And finally there is the stimulus of being able to unlock the mysteries of the heavens by a tool fashioned by one's own hand.

The framework within which the modern ATM works is very different from that of even 20 years ago. The relative expense of a commercial telescope has decreased to the point that many more people can afford to purchase good instruments. Yet there remains a motivation to build a good telescope oneself. For those of us afflicted with that motivation the same forces that changed the world of commercially made telescopes have changed the environment of amateur telescope making.

The existence of the Internet has allowed almost any producer of almost anything to "go global." Accordingly a producer of astronomical mirrors in Asia can ship them to Mexico, where they are combined with parts made in India and machined

R.L. Clark, *Amateur Telescope Making in the Internet Age*, Patrick Moore's
Practical Astronomy Series, DOI 10.1007/978-1-4419-6415-1_1,
© Springer Science+Business Media, LLC 2011

with things or equipment made in Turkey to be sold in the United States. A side effect of all this commerce is that overruns, components for discontinued models, small production runs, and used equipment can appear on the global market. After WWII the idea of "war surplus" became part of the typical American consciousness. We even came to realize that optical components might be available at our local surplus store. Who, of that age, didn't come into contact with a "tank prism"? How many of those things did they make? They still show up for sale.

There have always been shrewd business folks who could buy what was useless to one party and find a market that the original owner didn't imagine existed. These were, and still are, called "liquidators." The liquidators and war surplus dealers needed the talent and courage to make considerable investments with the hope of turning a profit.

Now the Internet allows anyone to play the game. There are secondary, tertiary, etc., markets – people who buy on eBay and re-sell stuff that someone didn't want in the first place. There are catalog stores whose presence is largely web. Many standard "brick and mortar" stores now have a catalog and a web page. And there are individuals making something in their basement and selling it worldwide. Some sources have no web presence but are, nevertheless, important and should be regarded just as if they were offerings on eBay. Those sources are "yard sales," particularly yard sale optical devices that may contain good components. For example, old 35 mm cameras often have lenses that make nice wide-angle eyepieces for fast (low f ratio) telescopes.

One way of classifying the huge array of optical stuff available on the Internet is according to the original, designed, purpose of the items. Some items were produced to be telescope parts. They may be items from production overruns or specific items produced for Internet distribution. The manufacturer-seller avoids the overhead costs of more standard distribution and is thus able to sell at a significant saving. Many of the parabolic mirrors available are in the later class.

Other items are surplus from non-telescope activities. It is among those items that the really low prices are available. Such prices are low in terms of original manufacturing cost and/or intended sale price. A fundamental question is always "Is this item suitable for use as part of a telescope and, if so, what part?" For example, items designed to deal with monochromatic light are not going to work well in white light application.

Hopefully this book will help in the sorting process. There is a lot of good stuff out there in both classes.

A very significant part of the content of this book deals with the construction details involved in assembling the various parts and pieces into useful telescopes. These construction techniques/tricks are not really dependent on the web, but without them you are unlikely to come up with a good instrument.

The non-optical parts and materials involved in telescope construction are just as important as the optical parts. In this book such materials are pretty much restricted to items available from local sources, such as a hardware store. Almost everything can be accomplished with a good collection of hand tools, but a table saw, power drill, and disk sander will help a lot. Most of the telescopes used as examples in this book were constructed as examples, specifically for the book. Plans and specifications were created. Then the instrument was built. In the cases where the plans needed to be changed in the light of reality the changes were made as the instrument was built.

Tools

There are several categories.

Absolutely Necessary

You probably already have these tools:

- Hand saw (back saw is good choice)
- Power hand drill
- Manual hand drill
- Several hammers
- Assorted screwdrivers
- Assorted wrenches
- Soldering iron

Very Helpful or Necessary for Particular Tasks

This depends on what you actually need to do.

- Mototool (Black & Decker calls it a "Wizard Rotary Tool")
- Table saw
- Various files
- Various chisels

Nice to Have

- Lathe
- Disk sander

Appendix I in this book contains a list of suppliers and a cross-reference table indexing the suppliers by the general kinds of items they provide.

This book is designed to allow a certain amount of jumping around. There are two reasons. First, you may find that some sections are particularly boring because you already know as much or more of that particular material as we are presenting. And second, some sections may be of little interest. It is entirely reasonable to put them off indefinitely or until you need the material.

By the same standard, there are a few sections that you either read or make sure you already know the material. These include the chapters on telescope types and design, the one on testing and quality, and the one on copy-scopes. You may decide not to actually build a copy-scope, but much of the material applies to the building/design of any type of telescope. The copy-scope chapter is simply a convenient place to discuss that material.

Several telescope projects, some described briefly below, are presented in detail. Look at the first sentence of this introduction, where we talk about the independent nature of telescope builders. Hopefully the projects presented in detail will be adapted by the reader, in keeping with the spirit of independence characteristic of the ATM community. In other words, don't just build what is described. Use ideas from it to build something better.

Dollar Store Telescope. All optical parts are from the dollar store or even your desk drawer. This is an uncorrected, super easy, and super cheap little telescope to illustrate basic principles and various aberrations. It will work and, depending on the accidental quality of the optics, may turn out to be better than anything Galileo had.

f5 6-inch Newtonian. Mirror from eBay, secondary from eBay, eyepiece from online catalog, focuser from online catalog, spider homemade, tube homemade wood. Cost about $90.

f9 6 inch Newtonian. Mirror homemade, secondary from eBay, eyepiece from online catalog, focuser from online catalog, spider homemade, tube homemade wood. cost about $70.

Small CopyScope with portable mounting. Objective lens from online catalog, eyepiece from yard sale camera, 1st surface mirror for star diagonal from eBay, tube homemade wood, projection lens from online catalog for solar projection. Cost about $35.

Large CopyScope. Objective lens from dumpster, eyepiece from online catalog, tube homemade wood, focuser from Orion catalog. Cost about $25.

Even Larger CopyScope. Objective lens from eBay, eyepiece from eBay, camera and focuser from Orion catalog, tube homemade wood. Cost about $60.

f11 80 mm Refractor. Objective lens from online catalog, focuser from Orion catalog, eyepiece from eBay, tube homemade PVC, finder objective from eBay (projector lens), finder eyepiece lenses from eBay, Cost about $60.

F17 4 inch Refractor with rotating star diagonal. Objective lens homemade doublet, tube homemade wood, 1st surface mirror for star diagonal from eBay, hardware for rotating star diagonal pieces of PVC. Cost about $70.

F6 12.5 inch Newtonian with rotating top. Mirror from eBay, secondary from eBay, spider from eBay, focuser from online catalog, eyepiece from eBay, rotation mechanism from slippery panel, tube homemade wood. Cost about $450.

Ergonomics

Ergonomics is the science of making things easy. A person with a doctorate in Operations Research and the job title "Human Factors Engineer" once explained that human factors engineering was the science according to which "Stop" buttons are colored red. Many pieces of equipment, not just telescopes, are constructed with very little consideration of just how the thing might be used. If something is difficult and anti-intuitive to handle it won't be used much.

The ergonomics of telescope design are often ignored and should not be. A telescope must be easy, comfortable, and natural to use. Many are none of these. The result is that such a telescope is used much less than it might be. If an eyepiece position requires that the user stand on tiptoe, at the top of a ladder, or in a pretzel position that telescope won't be used much. Here are a few ergonomic disasters that are easily avoided:

Insufficient focuser adjustment range	This forces a near-sighted or far-sighted user to use their eyeglasses to get a decent focus. They are, thus, forced back beyond correct eye-relief distance and get to see only a little, central circle of the field
Sticky adjustment bearings	Results in an attempt to make a slight adjustment in pointing to avoid movement, then an over-adjustment. A jerky movement that makes it hard to make small adjustments
Finder eyepiece position not consistent with main eyepiece position	Observer has to make a significant change in position – balance to go from locating the object to actually looking at it, for example, a straight through finder on a Newtonian telescope
A refractor not equipped with a star diagonal	Causes stiff necks unless the object is close to the horizon
Finder adjustment difficult	The finder and the main telescope are never properly aligned
Finder attachment not solid	Same as above
Telescope not balanced	Telescope won't stay on target
Unsafe stool, chair, or ladder	Broken bones

What's Not Included in this Book

Most construction details for Dobsonian telescopes have been excluded from this book. An exception is the example of the big Newtonian (12.5 inch). It has an extra Dobsonian-type mount for the rare instances when it may be moved. The reason for generally excluding Dobsonians is the fact that there are many excellent sources for this information. Many of those are in the form of web pages. The optical train is the same as a normal Newtonian and that is covered.

The entire area of astrophotography is pretty much untouched by this book. That field is very much in transition. The advent of digital techniques has made a very profound change in what can be done and how it can be done.

General Advice

You are embarking on a project to make a scientific instrument. The ability to purchase parts relieves you of a significant amount of work; exactly how much depends on what you chose to purchase and what you chose to build. However, it doesn't relieve you of the need for careful and precise design and assembly work. You are, after all, building a scientific instrument. It has to be pretty good to be of any use at all!

Suppose you have built a telescope whose objective has a focal length of 30 inches. A pretty likely target will be the Moon. The image of the Moon produced by the objective will be a bit less than 1/3 of an inch in diameter. All of the data – information that you are going to get – is pushed into a 1/3 of an inch circle. Suppose that you would like to see some level of detail on a little crater (say the little craterlets on the floor of the lunar crater Plato); they are 3 or 4 miles in diameter. The primary image of a 3-mile craterlet will be pushed into about 1/3,000 of an inch. Your eyepiece magnification is required to blow that up enough to allow you to detect at least a little of its structure. That is a pretty high demand on both the eyepiece and the objective. A good telescope with an aperture of 4 inch or more should pass that test.

Half the fun of amateur telescope building comes from building non-standard telescopes. Within the framework of freedom of design you will benefit by using fairly common fixtures and attachments. If something breaks or you lose a nut or bolt in some dark sky location (miles from anything), you are much more likely to get a replacement 1/4 × 20 wing-nut from the local hardware store than a 3/16 × 32 wing-nut.

Design your telescopes to be easily taken apart. That will facilitate moving them around, but, even more importantly, it will allow you to clean optical surfaces and help in re-using parts of one telescope in another. Don't worry about weight if you don't need to. There is no such thing as a telescope or mount that is too heavy unless you have to carry it or unless it is out of balance. If things can be disassembled into sections of moderate weight or if you can provide a permanent mount and shelter you should focus on planning for and building instruments that don't sway and vibrate in the slightest breeze.

Safety Tips

Telescope making is not normally considered to be a dangerous activity, like skydiving. It is unlikely that your insurance company is going to raise your rates because you are making telescopes. Nevertheless it is possible to get hurt at almost anything. Here are a few areas of potential injury.

Power tools are always dangerous. Drills, grinders, routers, lathes, and power sanders don't give you much time to get your hand out of the way. They just bite. Always turn the tool off and disconnect it when changing settings and/or blades. If you use telescope building as an excuse to purchase tools that you have always wanted such as a table saw be sure you read the manual and seek instruction if needed.

Stay scared or at least respectful. As long as you are scared you are unlikely to "dope-off" and get hurt.

Most injuries from tools such as table saws occur when you are at the end of a cut. You may be watching the blade and the pencil mark very carefully until you are at the last 2 inch of the cut. From there it seems like nothing can go wrong, so you stop paying attention to exactly where the blade is. That is just when something will go wrong!

Chapter 2

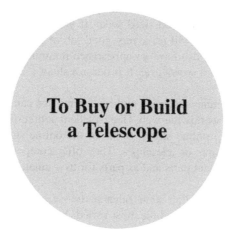

To Buy or Build
a Telescope

If you think you would like to own a telescope you are faced with the question: "Shall I just go out and buy a telescope or shall I build my own?" There are compelling arguments each way.

Buying is easier and quicker than building. Plus, over the past 20 or so years, costs for commercial telescopes have come way down. There are several reasons why this has happened. Part of it is the worldwide availability of components and assembled units. Part of it is just improved manufacturing technology. Part is competition.

However, it can be fun to build – if you have that kind of twisted mind! And, you are likely (though not guaranteed!) to learn something.

Plus, it is a nice ego trip to have folks look through something that you built and make WOW noises. If what you want is something different and/or unusual, you can do a web search on weird telescopes or go to Tom McMahon's page (Weird Telescopes), and then go from there. Note that none of those things is a first project.

Finally, you may find yourself building because you want something better than the standard manufactured scopes.

To be different a telescope does not need to be some sort of completely novel construct. You might choose to build simply to "buck the current trend." For example, before the beginning of the twentieth century, the most popular telescopes were refractors operating at speeds of about f15 and costing so much that average or even slightly rich folks could not purchase them. A 6-inch f15 refractor is a telescope at least 90 inch long that won't fit into most automobiles. These were/are very excellent instruments. If you ever get to try one you may well decide to make one. One of the example telescopes in this book is a f17 4-inch refractor instrument. Transportation requires a van or a pick-up truck, but boy does it produce nice planetary and lunar images. It has a very large "WOW" quotient.

R.L. Clark, *Amateur Telescope Making in the Internet Age*, Patrick Moore's
Practical Astronomy Series, DOI 10.1007/978-1-4419-6415-1_2,
© Springer Science+Business Media, LLC 2011

The current trend is toward reflectors operating at very fast f ratios. An f3.5 reflector with a diameter of 20 inch is on many modern-day Christmas lists. These are pretty expensive toys, so not a lot of those Christmas wishes will get fulfilled. You can get it into a good-sized sedan if you can deal with the 150–200 pounds of telescope, but it will do a very good job of pulling in galaxies known only by numbers. It will also have an appearance reminiscent of the Dardanelles gun. That will get a lot of wows even if it doesn't shoot a 2-foot stone across the Dardanelles.

Currently there are many telescope parts and items that can be used as telescope parts available on the worldwide-web. One excellent source is eBay, but there are many others, including online stores. Some of the online stores are operated by commercial manufacturers of telescopes. They offer components of their instruments, both as replacement parts and as parts for us – amateur telescope makers – to include in our projects.

Buying some parts, building some other parts – a combination approach – is the most practical route to go and the most varied. You might chose to buy optical parts and build the mechanical parts or go the other way, make your own optical components and buy mechanical parts, or a mixture of both.

A common first building effort is a copyscope. What is a copyscope used for?

A copyscope is often built to approximate a "richest field" telescope (RFT). The last chapter of the book *Amateur Telescoping Making Book Two* (ATM II) deals with the RFT concept. There are, in fact, two RFT concepts – the richest field telescope and a rich field telescope. It happens that both are covered by the abbreviation "RFT."

The intention behind both RFT concepts is to provide views with the most possible individual star images. There is sort of a formula for the design of such a telescope, and it is not a particularly difficult design to actually build. What you get is a telescope that, when pointed into the Milky Way, say in Cygnus, on a clear dark night will show you an eyeful of stars. Unless you have accidentally pointed at some special part of the sky with a lot of bright stars many of the stars you see will be perfect, faint, pinpricks of light.

The larger the aperture of a telescope the fainter the stars are that can be seen. That is because, theoretically, all the light that the telescope receives from a particular star gets concentrated into the image of the star. The bigger the objective, the more light, thus, the brighter the image. The following tables are rounded off and extracted from *ATM II*.

Diameter (inch)	Faintest magnitude
1	6.36
2.5	11
6.3	13
10	14
16	15

For example, if our telescope has an objective of 6.3 inch, we can expect to get down to magnitude 13. That assumes a clear, dark, sky.

Obviously, the higher the magnitude (the fainter the stars) the more stars there will be in a given area of the sky. Let's say you have an area of 1 square degree in the sky.

Top magnitude	Stars per square degree
8	1.6
9	3.5
10	9.8
11	25.4
12	61.7
13	141

Unfortunately the nature of the human eye limits how far we can go in increasing the aperture to increase the number of stars we can see in a single view. What stops us is the fact that, whatever light is gathered by a telescope, that light must all be pushed into a human eye if we are to be able to "see it." That light must be provided in a beam no wider than the pupil of the observer's eye. The pupil is no larger than 0.3 inch even if the observer is well "dark adapted" and reasonably young. As we age the ability to "dark adapt" becomes less.

The lower the magnifying power used the wider is the field of view, which means that lower powers will show more stars. There is a definite limit to how low we can go with the power, however. That "Limbo Stick" condition is imposed by the characteristics of the human eye. The 0.3-inch figure provides the limit.

The beam of light exiting from the eyepiece is called the "Ramsden disk." The diameter of the Ramsden disk is always the diameter of the objective divided by the magnifying power. Sometimes the Ramsden disk is called the "exit pupil." Thus a 6-inch objective used at a magnifying power of 22 gives a Ramsden disk of 0.273. That is close to our 0.3 figure for the eye, so that is almost all the light that the observer's eye can use. We could reduce the power to 21 and get a little improvement, i.e., 0.286. A little additional improvement from 22 power, i.e., 0.3, and we are done. The result, so far, is that lowering the power brings in more stars because it covers more area in the sky.

So, a 10-inch telescope operating at 33 power will give us the richest field available from any 10-inch telescopes. A 6-inch telescope operating at 22 power does it for 6-inch telescopes. A 2.5-inch telescope operating at 8 1/2 power does it for 2.5-inch telescopes.

The 2.5-inch design is the one that will show the absolute maximum number of stars in a single view. The specs to achieve that are a diameter of about 2 1/2 inch and a power of about 8 1/2. This is defined as the richest-field telescope. The other designs maximizing star counts for various diameters are called "rich field telescopes." There is a difference. For details, see S. L. Walkdens' article "Richest-Field Telescope" near the end of *Amateur Telescope Making Book II*.

On a good, dark, clear night you should see hundreds of stars any place you look in the Milky Way.

Many of the Messier objects will show up with great beauty in a RFT. The following have a lot of WOW! power.

M31	The Andromeda Galaxy	(Island galaxy)
M45	The Pleiades	(Open star cluster)
M42	The Orion Nebula	(Emission Nebula)
M44	The Beehive	(Open cluster)
M81	Island Galaxy in Big Dipper	(Galaxy)
M13	Hercules Cluster	(Globular cluster)

A nice trick for seeing the best of these objects is to locate them, get the focus really good, then move the telescope to the west just enough to get the object out of the view. Now just wait for the object to be brought into view by Earth's rotation. You will see more of the edges, etc., of the objects that way.

These are all "extended" objects; that is, they cover significant areas of the sky. The Andromeda Galaxy is "several Moons wide." Seeing these objects does not need much magnification power. It needs light-gathering power. At high or even medium power many of the neat "wow" objects won't even fit in the view.

A little Internet searching will get sky charts showing where to find these objects. If you do a web search on "copyscope" you will get lots of hits. In a test search the first 3 hits were perfectly satisfactory "how to make one" write-ups. Accordingly there is no need to go into a lot of detail here. Use the picture of it in this chapter, the ideas from the web presentations, or some combination of your own ideas and everyone else's.

On eBay we found three lenses that would work, two "very promising" and one good.

It is a little difficult to figure out what you are looking at under "telescope lens" on eBAY. You are looking for a lens with a diameter of at least 1 1/2 inch and a focal length at least 5 times its diameter. (That would make it an f5 lens.) Avoid f4 and f3 lenses. They can be very good, but good ones are very expensive and unlikely to be on eBay. The lens on this scope is 2-inch diameter and f8. (fl = 16 inch). The eBay folks don't seem to be able to differentiate between eyepieces and objective lenses. YOU ARE LOOKING FOR AN OBJECTIVE LENS. YOU SHOULD PAY NO MORE THAN $30 for a perfectly fine copyscope objective. American Science & Surplus has an f4 or so lens for $12.50. You probably would want to reduce its diameter with a black paper ring. Thus get it down to f5 or so. If you end up working with a lens with a low focal ratio you should try it at full aperture just to allow yourself to see the effect of coma and spherical aberration. Then put the black paper ring on to get the performance reasonable. A terrestrial telescope can supply perfectly satisfactory views with quite a lot of optical faults. A star image is a much stronger test of quality.

Don't buy a lens blank. This is not a lens; rather, it is one of the two or more pieces of glass from which a lens might be ground and polished.

A search for "telescope lens" yielded the following: (Fl refers to focal length.)

Fl	Dia	f ratio	Buy now price
Abt 20 inch	2 inch	f 10	$6.95

This would make a very satisfactory, small telescope. The aperture is a bit small for a RFT, but it would be easy to build and would make a nice first telescope. Also, it would be an excellent finder for your next, larger telescope.

Fl	Dia	f ratio	Buy now price
32 inch	2.44 inch	f9	$202

This is a camera lens and should work very well. A 4-inch fl camera lens used as an eyepiece would give you a really nice RFT. With a 3/4-inch fl standard eyepiece you would get about 50 power.

Fl	Dia	f ratio	Buy now price
72 inch	6 inch	f12	$630

This lens could be the heart of a really nice 6-inch refractor. It certainly is not an RFT. It would be useful as part of a planetary instrument. It would work very well as an only telescope, but just not as a "faint fuzzy" instrument. Keep in mind that you would get a telescope almost 7 feet long and not very lightweight. To operate at all well it would need a pretty hefty mounting. It would be a bit of a struggle on a train. Telescopes in this range are, generally, several thousand dollars.

Fl	Dia	f ratio	Buy now price
26 inch	2.36 inch	f11	$7

This would make a very satisfactory small telescope. The aperture is OK for a RFT, but you would need a 3-inch focal length eyepiece to get it down to RFT power. It would make a nice first telescope. It would also be an excellent finder for your next, larger telescope.

Fl	Dia	f ratio	Buy now price
36 inch	2.36 inch	f15	$15

Used very much like the one described above except that you could get a little more magnification.

There are lots of things you can use as an eyepiece. The lens from a 35 mm camera is likely to be good. Just make sure you get one from a camera with the shutter behind the lens, not between two parts of the lens. (Try yard sales for old film-type 35 mm). If the shutter happens to be stuck, so much the better – it's cheaper. Try not to spend more than $2 or $3.

You want a relatively low power eyepiece. Something with a focal length of 3/4 of an inch to 2 inch is good. A combination of a 2-inch focal length eyepiece with a 16 inch focal length objective lens gives 8 power.

A small eyepiece used with the illustrated copyscope when doing solar projection seems to be the same as the $7.50 Cinepar at American Science & Surplus. The eyepiece normally used for observing is a 50 mm (2-inch) fl lens from a 35 mm camera.

The telescope shown has an extra back with a tube sticking out of it just 1 1/4 inch inside in diameter. This allows the use of standard eyepieces.

You can also buy a reasonably good eyepiece on eBay, which usually designates the focal length in millimeters; so a 50 mm eyepiece is 2 inch. A 20 mm eyepiece is a little less than 1 inch.

Following is a little more information on eyepieces.

There are three "standard" diameter sizes for the tube:

1 1/4 inch (the most popular standard)
2 inch (better than 1 1/4 but more expensive)
0.956 inch

The last is the standard for microscopes, and you might want to try a microscope eyepiece on your telescope. It often works pretty well and is not uncommonly found on bargain-priced telescopes.

You need to be able to move the eyepiece in and out to focus. There is a plumbing fixture designed to fit the "trap" on a sink that has an inside diameter of 1 1/4 inch. This will give you a pretty good slip fit.

You can buy a focuser on eBay for about $20–$30. Orion makes a nice one for about $30. It is mostly plastic but works very nicely.

The magnification provided by a telescope is calculated by dividing the focal length of the objective by the focal length of the eyepiece. By changing eyepieces you can change the power.

A web search will produce many pictures – diagrams – of homemade telescopes. Here is a picture of one of the author's. Some suggestions regarding physical design are as follows:

- PVC pipe is useful for the tube.
- Plywood is nice. You can build a simple box if you like.
- Borrow parts of any design you like.
- Keep optics lined up.

To get a lens mounted on a piece of plywood for the front end you may get lucky and have a lens with a flange that makes mounting it easy. Otherwise consider using an automobile hose clamp on both sides of the lens.

You can, and most likely will, get ideas for improvements as you go. You are building something for your use, not for eternity.

The black box on the back end of the scope in Fig. 2.1 is called a "star diagonal." It allows you to avoid a lot of stiff neck problems when looking up.

The advantages of using a copyscope as a first project are that you can finish in not much more than a weekend, and also that it gives you something to use while you work on your second project. The usual second project is a 6-inch or 8-inch Newtonian reflector. You can use your copy-scope as a finder for your bigger scope. The Texerau book cited in the Appendix of this book will take you through the mirror making process in a step-by-step way.

Fig. 2.1 Example of a copyscope

If you don't want to invest a minimum of 40–60 h in a mirror you will find a pretty good selection on eBay. Most 6- and 8-inch mirrors are available for less than $100. They seem to be pretty decent quality, but buyer beware!. If you are going to make your own, here are some suggestions:

- Try to make it better than any you might buy.
- Don't try to save money on the blank or blanks. Use Pyrex for the mirror and get a thick blank.
- Build a Faucault machine a la Texerau and learn to use it. (Move the lamp box forward about an inch from his design position unless you have no nose.) The steel bar from an old dot matrix printer will work very well as a slide on the machine.

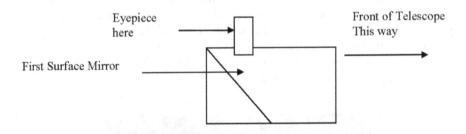

Fig. 2.2 Star diagonal

Fig. 2.3 A crop of home-built telescopes

A first surface mirror is one with the "silver" (actually aluminum) on the front side. Otherwise you will get double images. For this application, you don't need a super high quality one. Mirrors can be salvaged from old copy machines or purchased from numerous online sources.

No telescope is going to be very much fun to use if you don't have a solid way of pointing it and keeping it steady. You need a mount. If you have a reasonably good photographic tripod you can probably use that. The standard way to attach something like a camera or telescope is with a $1/4 \times 20$ inch thread machine screw, but this may not be steady enough. Mounts made from iron pipe work well. See www.stargazing.net/mas/scope3/htm or www.charm.net/~jriley/sky/mount1.html.

There are many other web sources of mounts. A good beginning is www.memphisastro.org/Mounts.html.

Figure 2.2 shows see what might happen to you if you get bitten by the ATM (amateur telescope making) bug!

Get involved with a local astronomy club. Some of them run mirror-making workshops that will be very helpful if you want to grind and polish your own mirror to make a reflecting telescope. You might even find a lens-building workshop. At least you will find a huge source of suggestions, help, and opportunities to share your interest and enthusiasm.

Chapter 3

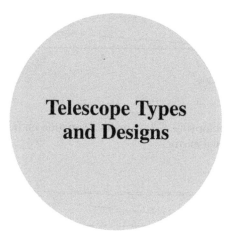

Telescope Types and Designs

In this chapter we examine the various telescope types from the point of view of an amateur telescope maker looking at the appropriate uses for each design type.

There is no "best design," but depending on the intended use, there are advantages associated with each.

Telescope Types

In its most simplistic description a telescope has an "objective" to form a "primary" image and an "eyepiece" that allows us to see the primary image. There are two basic varieties, reflector and refractor. The difference between the two types is only in what is used as an objective. Eyepieces are pretty universal.

Reflector Telescopes

A Newtonian reflector uses a mirror to form the primary image. This design is due to Isaac Newton. Hence the name Newtonian reflector (Fig. 3.1).

The Newtonian design is used by more amateur telescope users than any other design. This is true for both serious homemade 'scopes and commercial 'scopes. Dobsonian telescopes are Newtonians but on a special mount.

If the f ratio of a Newtonian is less than about f8 the objective (often called the primary) needs to take the form of a parabola rather than that of a sphere. The common approach to producing a parabola is to first make the sphere similar to the desired parabola, then "figure" it into a parabola. The difference is very small. For a 6-inch f8 mirror the difference is about 11.4 millionths of an inch.

R.L. Clark, *Amateur Telescope Making in the Internet Age*, Patrick Moore's
Practical Astronomy Series, DOI 10.1007/978-1-4419-6415-1_3,
© Springer Science+Business Media, LLC 2011

Fig. 3.1 Newtonian reflector

A Herschelian reflector (Fig. 3.2) is really a variation on the Newtonian design. It does not use a diagonal mirror.

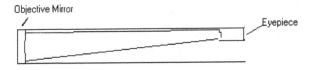

Fig. 3.2 Herschelian reflector

The Herschelian design has the advantage of not needing a secondary mirror, so it has only one optical surface. This makes it a bit more efficient, since you lose some light at each surface. In addition it has the advantage of not having an obstruction right in the middle of the tube. The diagonal blocks some light in a Newtonian. It has the major disadvantage of requiring that the objective mirror be set at an angle. This introduces some astigmatism into the system. If the f ratio of a Herschelian is below about 8 then stars will look like little lines. This is a symptom of serious astigmatism. If the f ratio is in the range of 15–20 or above the astigmatism will be unnoticeable, but that means that, for a 10-inch diameter mirror you will have a tube 150–200 inches long and the observer has to be at the upper end. This is certainly not an easy telescope to use. In any case the position of the observer's head leads to rising heat currents in the optical path, which leads to an unsteady image.

Some Herschelian telescopes have been built as "folded Herschelians." The idea is to first use surface mirrors to make the optical path traverse the length of the tube more than once. In that way the overall length of the telescope is reduced. An example telescope belonging to the author is about 8 feet long. Without the folds it would be 18 feet long. It has an f ratio of more than 50, so astigmatism is not a problem. It can achieve 200 power (about its limit) with an eyepiece that has focal length a little more than 1 inch and allows long, comfortable eye relief.

There are two additional designs that use only mirrors (except for the lenses in the eyepiece). A Cassegrain reflector has a parabolic mirror as an objective but uses a small convex mirror to shoot the image back at the objective and through a hole in the objective to reach the eyepiece. The Cassegrain (commonly called a Cass) has the advantage of offering a short tube relative to its focal length – f ratio. It is an excellent choice for planetary observations because the long focal length allows high magnification without a long tube (Fig. 3.3).

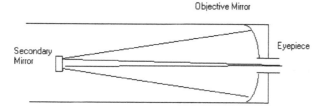

Fig. 3.3 A Cassegrain telescope

A Gregorian reflector is somewhat like a Cass but uses a concave secondary mirror. The focus is lengthened, but so is the tube (though not as much as is the focal length). The Gregorian also produces a "right side up" image. This is a small advantage, since who cares if an astronomical image is reversed. Also, we aren't going to use the telescope effectively on terrestrial (Earth-based) objects because lots of stray light will come in around the secondary and wash out our image.

Gregorians are very seldom made by amateurs or professionals. In high focal ratios they should be easier than a Cass and could be wonderful planetary instruments (Fig. 3.4).

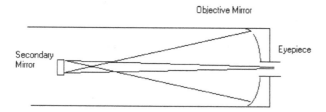

Fig. 3.4 A Gregorian telescope

Refractor Telescopes

A refractor uses a lens to form the primary image. A refractor looks like a telescope. A reflector looks like a piece of pipe with a bump on the side (Fig. 3.5).

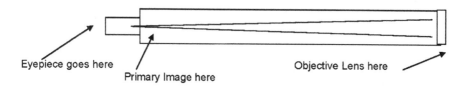

Fig. 3.5 Refractor telescope

One of the effects of the current web availability of optical components is that the typical "copyscope" refractor can replace the small Newtonian design as your first serious telescope. Making a refractor is pretty simple except for making the objective lens. *That part is entirely do-able but not simple at all.* If you can get your hands on a suitable objective lens the rest is easy. In fact that is the easiest way to build a telescope. If you use a lens that was not originally intended for a telescope you still may get a pretty good telescope.

Telescopes with such salvaged lenses are often called "copyscopes" because junked office copy machines are often the source for the lenses. As a matter of fact a copier lens is designed for about "one to one" projection. This is not quite ideal but will work. The job of a telescope objective is to accomplish an "infinity to one" projection. A good-sized projector lens or even a really big camera lens may be better.

A copyscope is different from other refractor telescopes only because the salvage lenses from copy machines, projectors, cameras, etc., tend to have fairly large diameters and relatively short focal lengths, producing a focal ratio in the f4–f8 range. That makes them approximately the "richest field" specification. Otherwise there is no real difference. A copyscope might be defined as just a refractor made with an objective lens salvaged from a non-telescope application. Two copyscopes and two refractors that are not copyscopes are presented as examples in this book.

A good copyscope makes an excellent first telescope-making project.

There are several designs that use mirrors as primary image sources but which also use lens components to adjust and correct the image. These are the so-called Schmidt-Cassegrains, or catadioptric telescopes or Maksutov telescopes. Although these designs can be built by an advanced amateur they are well beyond the scope of this book.

There is a design called Ritchey-Cretien that uses two mirrors and no lenses other than those in the eyepiece. The mirrors are hyperbolic. Additional mirror-based designs are the Schiefspiegler and the Yolo. These also can be built by an advanced amateur but are well beyond the scope of this book.

Telescope Design Considerations

The design that you choose for your home-built telescope should depend on what you intend to do with the telescope. There is no such thing as an "all-purpose" telescope. Some of the commercial manufacturers attempt to get as close to an "all purpose" instrument as they can. This effort makes absolute sense from a marketing point of view but may not be so good for the buyer. As a builder you have the wonderful freedom to build an instrument for a restricted purpose. If you want another telescope you can just build a second instrument.

Let's now look at the criteria you should consider in choosing a design.

First, what type of object do you want to optimize your design for? Some object types, such as nebula and galaxies, are best seen with an instrument with as much light grasp as you can get, but usually you don't need much magnification. Magnification in the 20× to 40× range is plenty for most. These are the objects that are often called "faint fuzzies." A wide aperture, low power instrument is ideal. The typical "copyscope" (see next chapter) designed along toward the specifications of a "richest field" telescope will show many of these objects. There are exceptions, such as the ring nebular (M 57), which needs at least 60–100 power.

Following is a chart of the angles subtended by the 110 Messier objects. For comparison, consider that the Moon (obviously also the Sun) subtends about 1/2 degree.

The second column in the chart indicates the type object. GLOB means globular cluster. NEB means nebular. OP means open cluster. The third column contains the angular size of the object in minutes of arc. Keep in mind the fact that the Moon subtends about 30 min of arc. The average of the Messier objects is about 20 min of arc, and 20 min of arc is 2/3 of a Moon. Even if we exclude the three biggest we get about 1/2 of a Moon. Except for the smallest (in terms of angle subtended) Messier objects they subtend angles that would make them naked eye objects if they were bright enough. What we really need is light-gathering capability, not great magnification. Most of the Messier objects deserve the term "faint fuzzies."

M1	NEB	6
M2	GLOB	16
M3	GLOB	19
M4	GLOB	35
M6	OP	20
M7	OP	80
M8	NEB	90
M9	GLOB	11
M11	GLOB	13
M12	GLOB	14
M13	GLOB	21
M14	GLOB	11
M15	GLOB	18
M16	OP + NEB	120
M17	OP + NEB	40
M18	OP	5
M19	GLOB	14
M20	OP + NEB	28
M21	OP	15
M22	GLOB	33
M23	OP	30

M24	OP	7
M25	OP	30
M26	OP	8
M27	NEB	8
M28	GLOB	10
M29	OP	10
M30	GLOB	12
M31	GAL	180
M32	GAL	8.7
M33	GAL	71
M34	OP	25
M35	OP	25
M36	OP	10
M37	OP	15
M38	OP	15
M39	OP	30
M40	DBL ST	0.8
M41	OP	40
M42	NEB	90
M43	NEB	20
M44	OP	72
M45	OP	120
M46	OP	20
M47	OP	25
M48	OP	30
M49	GAL	10.2
M50	OP	15
M51	GAL	11.2
M52	OP	16
M53	GLOB	13
M54	GLOB	12
M55	GLOB	19
M56	GLOB	7
M57	NEB	1.2
M58	GAL	5.9
M59	GAL	5.4
M60	GAL	7.4
M61	GAL	6.5
M62	GLOB	11
M63	GAL	12.6
M64	GAL	10
M65	GAL	9.8
M66	GAL	9.1
M67	OP	25
M68	GLOB	11
M69	GLOB	10
M70	GLOB	8

M71	GLOB	7.2
M72	GLOB	6
M73	Asterism	1.4
M74	GAL	10.5
M75	GLOB	7
M76	NEB	1.1
M77	GAL	7.1
M78	NEB	8
M79	GLOB	6
M80	GLOB	9
M81	GAL	26.9
M82	GAL	11.2
M83	GAL	12.9
M84	GAL	6.5
M85	GAL	7.1
M86	GAL	8.9
M87	GAL	8.3
M88	GAL	6.9
M89	GAL	5.1
M90	GAL	9.5
M91	GAL	5.4
M92	GLOB	14
M93	OP	10
M94	GAL	11.2
M95	GAL	7.4
M96	GAL	7.6
M97	NEB	2.2
M98	GAL	9.8
M99	GAL	5.4
M100	GAL	7.4
M101	GAL	28.8
M102	DUPE 101	
M103	OP	6
M104	GAL	8.7
M105	GAL	5.4
M106	GAL	18.6
M107	GLOB	13
M108	GAL	9.7
M109	GAL	7.6
M110	GAL	21.9
	Average	19.85

Note that most of these objects become less distinct as we observe them toward their edges, so the angles subtended are somewhat optimistic. Also note that, in cases where the shape of the object is distinctly not round, as is the case with galaxies, we have used the larger dimension.

A 6- or 8-inch Newtonian reflector is often thought of as a first serious telescope. A few years ago most of us (ATMs) ground and polished our own mirrors. The difference between a 6-inch and an 8-inch mirror is significant in terms of light-gathering capability (think faint fuzzies), but it is also significant in terms of grinding and polishing work. Most of us did a 6-inch as our first serious effort. You can, of course, go ahead and do your own grinding and polishing, but unless you want an extreme f ratio, give serious consideration to offerings on eBay. At the time this was written there were:

2	8-inch f4	$80; these were only 3/4 inch thick, so a little thin but OK if mounted well
1	6-inch f5	$50; this is only 5/8 inch thick, so also a little thin but ok if mounted well
1	8-inch f7	Current bid was $32. It had a few fine scratches. It probably sold for $50–$80. It is 1 3/8 inches thick so more stable than the previous 8-inch
1	12.5-inch f6	Buy It Now price was $375. It was Pyrex and 2 inches thick, so good and stiff and will cold adapt fairly quickly. Probably more than you want to mess with as a first effort, though. At f6 you will have an eyepiece located about 70 inches from the bottom end of the tube. If the telescope is pointing straight up and the bottom clears the ground by 2 or 3 inches it may require a ladder or stool to get to the eyepiece. A mirror like this is the basis for the telescope described in Chap. 10 of this book

The first two items were new, so they probably had good coatings. The last two were used. The seller said that the coatings were fine. The buyers might have needed to get them re-coated, though any fine scratches were probably harmless.

The raw materials for any of the listed mirrors would cost about half the cost of a finished mirror. You would then have to get your finished mirror coated. That might well use up much of the other half.

Now as to a second criteria: to what extent is portability a factor?

If you have an area with a decent sky that you can use on a permanent basis, and you don't intend to move your equipment, you may be able to ignore the weight. That can be a very important plus, because weight tends to mean steadiness. Concrete is an excellent building material, but it doesn't lead to portability. An instrument on a concrete mount with a slide-off roof allows you to have a very steady view and precludes you from lugging the telescope and mount from your house/apartment to the observing site. A solid mounting that can stand the weather together with some sort of weatherproof outdoor structure for the telescope can save a lot of lugging.

On the other hand, if you expect to travel to a dark-sky observing site portability is critical. Things have to be light enough to carry easily, and they have to be easy to set up. Remember – you may be able to set up before it gets dark, but you are sure to be tearing down in the dark. An obvious way to achieve the lightweight requirement is to design things so that they can be disassembled into several pieces and reassembled on site. Keep it simple. Equipment that you can carry around in your yard during the day may be a real problem in the middle of a cold night in the dark, in a plowed field, or down a dirt road at 3:00 a.m. after you have been observing all night. Be very conservative.

The Dobsonian mount design and its variations are pretty well adapted for portability. Since the first two (relatively thin) eBay offerings would be lightweight, they would be very appropriate for a portable Dobsonian, telescope.

Two other criteria for selecting a design are: How much you are willing to spend, and how much work you are willing to do? These play off against each other. An obvious way to save money is to do it yourself. Most builders who chose to buy components rather than make them chose to buy optical components such as main mirrors for reflectors and make other components. Others want the experience of grinding and polishing, so they may make the mirror but purchase other parts.

Two last criteria: What tools are available, and what parts or materials do you already have?

Most people who contemplate building a telescope are the same folks who build other things. They are likely to have a reasonable collection of tools, but there are always other tools or attachments that these builders would like to have. Perhaps building a telescope can be parlayed into buying a table saw. If you have parts from some damaged telescope or someone's previous effort to build one, this may influence your choice.

A Real Simple Telescope

A book on telescope building should get the reader building a telescope as early in the reading process as possible. Toward that end here is about the easiest telescope that can be built. You may already have all the required parts, and, if you don't, a trip to the local "dollar store" should fill in the gaps. An alternate name for this little telescope is a "dollarscope".

You can put this thing together in a few minutes plus some glue drying time. You will end up with a working telescope at least comparable to the one used by Galileo. It will provide a basis for some interesting experiments and allow you to see the effects of the most common optical aberrations. A very easy alteration will convert it into something that many, otherwise smart, folks will claim can't exist.

One of the important uses for this little telescope is to give you some practical experience in identifying the various optical aberrations that people talk about but often only vaguely understand. It should also provide palpable proof that the quality demands on an astronomical telescope are far greater than those on a terrestrial instrument. What may seem OK when looking at wild birds can be pretty useless on a celestial object.

For this little telescope you need to acquire two magnifying glasses, one to be used as an objective lens and the other as an eyepiece. The dollar store in your area is a likely source. Your own top desk drawer may supply just what you need. Quality is not very important. The lens to become the objective lens should be reasonably large – 1 1/2 inches or more in diameter is appropriate. The magnifiers sold to aid reading fine print are good. They should be able to be used at a convenient distance from to what is being magnified – 5 inches or more. This is just a way of saying that the focal length should be 5 inches or more. What you get is a rather weak magnifier. The lens that will become the eyepiece should be smaller and stronger. One inch diameter is plenty, but it needs to be held pretty close to its subject. One or two inches is about right.

The magnifying power of this telescope will be the result of dividing the objective focal length by the eyepiece focal length. The combination of a 5-inch focal length objective with a 1-inch eyepiece will give you a 5 power telescope. Those two lenses would need to be spaced the sum of their focal lengths, $5 + 1 = 6$ inches apart. This is the "sum" distance and establishes the separation for perfect focus on an object at infinity. Different focal lengths give different answers. A later chapter discusses the adjustment appropriate in the separation of the eyepiece from the objective. A total adjustment of 1 inch around the 6-inch figure ought to be plenty in this case.

You can check the distance by holding the smaller lens close to your eye and holding the objective lens up about 6 inches in front of it. With a very small amount of adjustment of the objective, in and out relative to the eyepiece lens, you should be able to see an inverted image of a distant object with the lenses separated by the sum distance. For objects closer you will need to increase the separation a little.

To actually construct the telescope you need to make two tubes and a way of sliding one inside another. Use the 5- and 1-inch figures as examples: Make or get a tube 5 inches long and the same inside diameter as the objective lens. If you can't scrounge a tube (food can, mailing tube, etc.) you can make one by rolling up enough cardboard from file folders to get a wall thickness of three or four layers. Slather a good bit of glue between the layers and use scotch tape, paper clips and/or rubber bands to hold it together until the glue dries. Do the same thing for the eyepiece tube. It needs to be no more than 3 inches long.

You need a way to attach the lenses to tubes and a way of sliding the smaller tube into the larger one to allow you to focus the telescope. You can either make a couple of doughnuts from thick cardboard and glue them in place or use the rolled up strips technique to make a rather substantial doughnut. You can even use scotch tape. Just don't get a lot of tape on the surfaces of the lenses and get the lenses fairly square to the tubes.

Figure 3.6 shows a dollar telescope using some mailing tubes and a glued-up tube made from a couple of old file folders, in case you can't find ready-made tubes that will work. Don't worry about lumps in a glued-up tube. The whole thing will be flexible enough that you will be able to push it into shape. If you choose to add more layers you will end up with a pretty decent tube. This is probably how the tubes for Galileo's telescopes were made. There was a shortage of mailing tubes available in 1608 or so.

The lenses in the mailing tube design are secured in the cardboard tubes between rings made from the same tube. To make a ring in a tube cut it out from the end of a scrap tube. Cut a piece out of the ring so it can be compressed to fit inside the tube. You need to remove about 4 times the thickness of the material from the ring. Apply

Fig. 3.6 "Dollar store" telescope from mailing tubes and a file-folder tube

plenty of glue and slip it into the main tube where you want it. You may need to do a couple of these to reduce the diameter enough so that you have a step for the lens to rest on. Now try it out!

As said earlier, part of the reason for building this telescope is to encourage experiments that can illustrate various optical problems. To get the most out of this set of experiments you will, probably spend several hours playing with them. Try the experiments described below.

1. Look at a fairly bright electric light at least 20 feet away. By making slight adjustments with the sliding tube, try to get as sharp a focus as you can. Now go to a position just a little out of focus. Then go a little out of focus the other way. This should involve sliding the tube only about an 1/8 of an inch each way.

   Look for rainbow colored fringes around the image of the bright light. These will go from red to blue as you go from inside of focus to outside of focus. The problem is chromatic aberration, caused by the fact that glass bends light slightly more or less depending on the color of the light. The white light you are looking at is broken down into its various color components by the objective lens acting as a prism. It is coming to a focus at slightly different distances behind the objective so you can't find a position for the eyepiece that gives a really sharp image. Either the blue is well focused and the red is a little out of focus or the red is sharp and the blue is fuzzy.

Now try looking through some sort of a color filter. Even a candy wrapper will work if it isn't too messed up with candy. You should be able to get a sharper image because you are allowing only a small range of colors. You can focus these (for example, the reds) and exclude the out of focus blue.

Chromatic aberration can be minimized by using two different kinds of glass in the same lens. Normally the two disks are called the crown element and the flint element. The effect is to "fold" the red light and the blue light together. Such lenses are called "doublets," and the effect is to make the lens an "achromat." Most lenses used in serious telescopes are achromats.

2. Try your telescope out on a bright star. Now that you know what chromatic aberration looks like, you will see it in a star image. Try it with the filter. The image should become smaller. Ideally the image of a star should approximate a true point. Failure to provide tight focus is obvious with a star image and, when it is the result of chromatic aberration, it is obvious with any bright, white object. Although it may not be so obvious with an extended object such as the Moon, it is equally damaging. The same effect of a point being "soft" means that the multitude of points that make up an extended image are fuzzy. It may not be very apparent, but the effect is to significantly reduce the resolution offered by the image.

Get as good a focus as you can on a star in the center of the field of view. Without changing anything look at stars near the edge of the field. They may look like birds with two wings, flying toward you or away from you. They also may look like comets with their heads pointing at the center of the field and the tails pointing outward. Most likely they will take on some of both distortions and look like smudges. The first aberration (the birds) is called "spherical aberration." The second (comets) is called coma. Coma is not named for comet, but both words come from a Greek word for hair. In a later chapter the various aberrations will be discussed in more detail. For the moment there isn't much that we can do about either coma or spherical aberration.

One partial solution is to increase the focal ratio of the telescope by reducing the diameter of the objective lens. If our objective lens is 2 inches in diameter and has a focal length of 6 inches we have built a f3 telescope. That is a very low focal ratio and would be regarded as very "fast" if it were made from truly excellent lenses. It certainly wasn't, so it is an invitation to a variety of aberrations. If you make a mask to put over the objective with a hole cut in the middle to reduce the effective diameter the visual images may improve drastically, but their brightness will be very much decreased. An important reason for having an astronomical telescope is to make very dim objects visible, so reducing the brightness somewhat defeats our purpose. It is worth keeping in mind that the classic refractors of the nineteenth and first half of the twentieth century were generally built at about f15. Newtonian reflectors of the same time period were also much slower than their more modern counterparts are. The faster instruments of today represent, in part, a concession to portability.

3. Get the lenses out of alignment. Do something to get the objective lens out of parallel with the eyepiece lens. You may be able to squeeze the side of a cardboard tube to get one side tilted in towards the eyepiece by about 1/4 of an inch. You may have to actually remove the objective lens and put it back with a good tilt. Even more than 1/4 inch is fine. We are giving your telescope a good case of astigmatism, so we might as well make sure it is easily detectable.

Now try it on brighter and fainter stars. Move back and forth between inside of focus to outside. You will see the images transform from little straight lines in one direction to little straight lines in a direction perpendicular to the first. Between (at the "best focus" position) the image will be a messy compromise between the two lines.

Astigmatism in telescopes is generally caused by failure to have things lined up correctly. Accordingly, it is an optical aberration that we can do something about. If we are very careful about making sure that optical elements are square to the optical axis we will avoid the aberration.

While you still have your telescope pretty grotesquely astigmatic try it on an extended object such as the Moon. You will get a soft image, but as you go in and out through focus the softness will be seen to be directional. The directions will correspond to the orientation of the little lines you saw with the star images. That kind of softness might not be so obvious if you had not seen it in the star images.

Astigmatism is, by far, the most common of the aberrations that cause telescopes to fail. Astigmatism can also be a defect of the human eye. If your eyeglass prescription has a term for "cylinder" your glasses are correcting for your astigmatism. A later chapter will provide ideas on how to build the correction into an eyepiece.

An alternative to the use of a big magnifying glass as an objective lens is to use a lens intended for eyeglasses. What you want is a lens intended for correction of rather mild farsightedness. A two-diopter lens will have a focal length of about 15 inches. A one-diopter lens will have a focal length of about 30 inches.

The relationship between diopters and focal length is the following: The power of a lens in diopters is the reciprocal of its focal length measured in meters.

Eyeglass lenses come from the manufacturer round and about 1 1/2 inches diameter. They are described by their power measured in diopters. The eyeglass store grinds the edges of them down to fit the style frames you choose.

A local eyeglass store should be able to supply a two or one diopter lens quite cheaply. A discarded pair of reading glasses may supply an appropriate lens, but it will have been shaped to fit someone's choice of frames and so be a little awkward to attach to a tube.

For the eyeglass version of the dollar telescope you will need a longer tube. You may be able to get a workable estimate of just how long by measuring the focal length imaging the Sun. The distance at which you get the smallest image is the focal length. You can calculate the magnifying power of the telescope by dividing the objective focal length by the eyepiece focal length.

The eyeglass version will have significant chromatic aberration but less spherical aberration and less coma. Those improvements follow from the fact that the eyeglass version is slower. It will have an f ratio of 10–20, depending on if you use a one or two diopter lens. The same experiment (twisting the objective lens) to give it a good, solid dose of astigmatism will work. Mailing tubes, paper towel tubes, or even pieces of pipe can serve to hold the whole thing together.

4. Let's now convert what you have to a one lens version of our telescope. Just remove the eyepiece and replace it with a pinhole. To use this telescope hold the pinhole as close to your eye as you can with comfort. You will note the following:

   • As you move change the distance between the pinhole eyepiece and the objective lens the magnifying power changes. The closer you get to having the pinhole actually at the focal point of the lens the greater the magnification is provided and the poorer the image becomes.
   • The smaller the pinhole the better the image quality. It never gets very good.
   • The smaller the pinhole the brighter the object has to be to get an image at all.

If you repeat the previous experiments you will be able to appreciate the fact that most of the aberrations came from the objective lens.

It should be pretty obvious that the "One Lens Telescope" is unlikely to become one of the more popular telescope designs. But it does prove that it can be done!

Chapter 4

Build a Good Copyscope

The name "copyscope" comes from the fact that many of these devices were made using the glass from old copier machines. A good copyscope makes an excellent first serious telescope-making project. If you chose the components carefully you will have a rich-field instrument capable of reaching many extended objects and some wonderful star-fields. In this chapter three scopes are described.

What you end up with when you build a copyscope depends on both how carefully you build it and what sort of a lens you use as the objective lens. A telescope probably qualifies as a copyscope if the objective lens was initially intended for some use other than as a telescope objective. That provides a very large range of possible telescopes, ranging from small, classic richest field designs through some pretty hefty examples. In this chapter we will present three different copyscopes then discuss some general principles applying to all. If you build a copyscope it is very unlikely that you will begin with an objective lens matching any one of our three examples. For that matter, why should you want to? The examples span enough variety to provide ideas no matter what you end-up with as your starting point.

There are roughly three kinds of copyscopes – small, medium, and large.

The Small Copyscope

Optical Parts

Objective Lens

Diameter = 2.5 inches
Focal Length = 12 inches

R.L. Clark, *Amateur Telescope Making in the Internet Age*, Patrick Moore's
Practical Astronomy Series, DOI 10.1007/978-1-4419-6415-1_4,
© Springer Science+Business Media, LLC 2011

This lens has a diaphragm. That indicates that it was intended for use in a camera. Web purchase about $25.

Eyepiece

From a 35 mm camera. The lens has a coarse but very smooth screw that makes a fine focuser. That is what the screw was used for when the lens was in its first incarnation as a camera lens.

Focuser

Part of the camera lens for the normal eyepiece. A sliding tube for 1 1/4 inch eyepieces. One of the most frequent uses for this telescope is for solar projection.

Mechanical Parts

Tube

Built from wood with plywood front and back. The back of the tube can accommodate a back plate with a star diagonal or an alternate back plate for 1 1/4 inch eyepieces. It could also accommodate a camera back.

Mounting

This telescope was given its own little mounting that is, essentially, part of the telescope. See Fig. 2.1 in Chapter 2. The scope and mounting are intended to be used on a table. A picnic table is ideal! The base swivels on a "Lazy Susan." The pivots are 1/4 × 20 machine screws with the heads sawed off. The threads end about 5/8 inch from the head, so there is a little unthreaded part suitable as a bearing. The telescope can be locked into position by pulling tight a nut attached to the little aluminum handle visable in Fig. 2.1.

Finder

None
Figure 2.1 in Chap. 2 illustrates the small copyscope.

The Medium Copyscope

Optical Parts

Objective Lens

Diameter = 4 1/2 inches
Effective Focal Length = 18 inches
Back Focal Length about 17 inches
Marked: Charles Beveler Company, East Orange NJ Series III

This lens was probably used in a large (auditorium size) projector. Its source was a dumpster in Fig. 4.1.

Eyepiece

From a 35 mm camera or any standard 1 1/4 inch eyepiece.

Focuser

This was an eBay purchase. Cost $15 plus $9 shipping.

Mechanical Parts

Tube

Built from wood with plywood front and back.

Fig. 4.1 Medium Copyscope

This is a straight through design. It has an entirely removable back plate that could accommodate a star diagonal or camera back.

Mounting

The comfortable altazimuth mount described in a later chapter was designed for this telescope. The telescope has a flat plate to match the bottom of the cradle on that mount.

Finder

Peep sight.

The Large Copyscope

Optical Parts

Objective Lens

Diameter = 5.7 inchesEffective focal length = 36 inches
Back focal length about 20 inches
Marked Booth Telephoto F6.5
Weight about 30 pounds

The barrel on this thing is 11 inches long and is pretty well filled with glass. Before someone left it out in the rain or dropped it in a pond it was a pretty expensive item.

This lens has a diaphragm, which indicates that it was intended for a camera. The word telephoto probably told us that anyway.

This lens was acquired by a friend of the author from an eBay seller. It had been exposed to water, apparently dirty water. The water had dried on internal lens surfaces and rusted the threads that would allow its disassembly for cleaning. The seller had despaired of getting it apart and sold it very cheaply. The authors' friend came to the same conclusion and gave it away. The author applied lots of WD40^{TM} and it drifted apart over a period of several days.

What is drifting? Often parts that have rusted together or otherwise become frozen, one to another, can be separated by the application of relatively small force but applied very sharply if a small notch is cut into one of the parts and the bit of a dull chisel is inserted in the notch so that a sharp rap on the chisel urges the piece in the correct direction. If you do that, you may get a little motion in the frozen

Fig. 4.2 The large copyscope. Note the dowels sticking out of the sides of the cradle at the balance point. They allow mounting on the PVC-concrete mount. You can see the results of drifting the lens apart at the front, to the right of the lens cell. The round item to the left of the telescope is a salvaged piece of tubing that fits the front of the lens cell and serves to protect the front of the lens from stray light

connection. If you have applied lots of penetrating oil and apply yet more once the motion has begun you may be able to unscrew the frozen parts. You may, also, break the glass in the lens, so this is something of an "if all else fails" solution. Drifting is a technique more often used on severely rusted farm machinery than on lens cells.

You may be able to see the notch in the front ring of the lens in Fig. 4.2.

Focuser

From eBay; about $15. Manufactured for Orion Corp.

Eyepiece

Objective lens from a 35-mm camera mounted on a 1 1/4 inch tube.

Mechanical Parts

Tube

Built from wood with plywood front and back. See the description of wooden tubes in the chapter on tubes. This is a 16-sided tube made from pine and sanded to almost round.

The objective lens had a ring with ten holes appropriate for screws to hold it on the front piece of plywood. The tube was built according to the ideas presented in the chapter on tubes. A 6-inch piece of PVC tubing would probably have worked as well.

Mounting

Attached to a piece of 3/8-inch plywood to fit the universal mount described in the mountings chapter. Note that, due to the significant weight of the objective lens, the balance point of the telescope is at the front end. It is also equipped with dowels so it can be mounted on the PVC-concrete mounting or the tripod used for the 4-inch f17 telescope.

Finder

Peep sight.

How to Make Your Copyscope

Now that we have seen three copyscope designs we can begin to plan your version. The cases that we have looked at should suggest a few rules and plenty of construction ideas. Some useful ideas can be drawn, also, from the 80 mm refractor described in a later chapter.

Select and acquire your objective lens. Much of the remainder of the copyscope will be designed around the objective lens, so you need to have it "in-hand" prior to making other decisions.

Focal Length

The length of your telescope is determined by the objective lens. Do the initial design work independently of the decision to use a star diagonal or not. Your first step has to be to determine the distance between the objective lens and the eyepiece. Once that distance is known you can leave it as a straight line or fold it at a right angle to accommodate the star diagonal. In most, but not all, cases the objective lens forms a primary image placed as far back from the lens as the focal length of the lens. There are some cases in which the primary image pops up significantly closer to the objective lens than the focal length. The result is that we have to think about two versions of focal length:

EFL   Effective focal length
BFL   Distance between the rear-most glass in the objective and the primary image

In the case of simple lenses and most other cases the EFL and BFL are rather close. The difference is likely to be about half the thickness of the lens itself. Some camera lenses are compound with parts of the lens considerably separate from other parts. The Dolmeyer designs, for example, are made from four pieces of glass, one pair at the front and a pair at the back (in this case the two pairs are alike). There is a couple of inches of air between the two pair. In one example the back focal length is 12 inches and the effective focal length is 16 inches. We use the BFL to determine where the image is to be found and the EFL to calculate the magnifying power of a telescope resulting from using the objective lens with a particular eyepiece. In other words, the BFL is a construction criteria and the EFL is a performance criteria.

A little algebra now. Arithmetic can give us the effective focal length if we suspect that it is different from the back focal length. Remember it is the BFL that we need to build our telescope. All we can get from knowing the EFL is the resulting magnifying power and the focal ratio of the lens.

For a simple lens the following formula applies:

$$1/d_1 + 1/d_2 = 1/f$$

If we set the lens between a light bulb and a white screen or piece of paper and arrange them so that an image of the light bulb was projected on the paper, then d_1 would be the distance between the lens and the light and d_2 would be the distance between the lens and the image. Then f would be the focal length of the lens. Obviously d_1 and d_2 are interchangeable. Notice that if you fool with the spacing to get $d_1 = d_2$ you get: $1/d_1 + 1/d_1 = 1/f$ or $2/d_1 = 1/f$ or $d_1 = 2$ f. That gives you a neat way of getting a focal length for a simple lens on a dark day when you can't get a solar image. Just get a focus, add the reciprocals of the two distances, and you get the reciprocal of the focal length.

You can calculate focal length or, more precisely, effective focal length by measuring image sizes. If we indicate the height of an object as h_1 and the height of its image as h_2 then the image heights will be proportional to their distances from the lens. That is:

$$d_1/d_2 = h_1/h_2$$

Measure one of the two distances, say d_1, and measure both h values. By using the relationship above (the second equation) you can solve for d_2. If the EFL is significantly larger than the BFL your derived value for d_2 will be significantly larger than the actual d_2. Plug the derived d_2 into the first equation and solve for the EFL.

Alignment

For any telescope it is very important that its optical parts be properly aligned. Bad alignment can't help but have a very negative effect on performance.

There are two ways for a telescope to be out of line. If a lens or mirror is shifted off the optical axis but still oriented at right angles to the optical axis the effect will be that the instrument is using an off-axis part of the image. If the offending lens or mirror is tilted with regard to the optical axis the effect will be to introduce astigmatism into the image. In most situations a telescope can tolerate shift much better than tilt. There is, of course, no reason why you can't make two errors and have both problems at the same time.

The effect of a centering error (that is what a displacement amounts to) is to cause the telescope to use a peripheral part of the primary image, which is much more prone to coma than the central part of the image. Stars will look like miniature comets. Most telescopes can be expected to show some, hopefully a small, amount of coma. If the problem has not been exaggerated by a centering error the little comets will have their heads pointing toward the center of the image and their tails radiating outward. If centering problems are present the heads will point someplace other than the center. Hopefully, the apparent center will be somewhere in the field of view, but in wild cases it may be entirely outside the field.

The effect of tilt (astigmatism) is to make images perform as if the offending optical component (lens or mirror) had a cylindrical aspect to its shape. When you attempt to find optimal focus the star images will never come to a good point. If you position the eyepiece just inside (toward the objective) of best focus you will see a short line rather than a slightly blurry version of a point. If you move to a position outside (away from the objective) you will see a short line at right angles to the inside version. The best focus position (between the inside and outside positions) will give a slightly blurred image or, in extreme cases, a + or an ×.

Both problems can be cured by adjustments in the alignment of optical parts. Professional and amateur built instruments plus reflectors and refractors should be equipped with ways (usually screws) to make the necessary adjustments. That is what the great collimation mystery is all about.

In a refractor telescope it is possible that astigmatism caused by tilt of the objective lens may not be present at all. That is, if you were very conscientious about how the objective lens was mounted it may actually be square to the tube. A good steel carpenter's square is indispensable for making sure that the front of the tube is really square to the sides.

In any scope the most likely problem is astigmatism. Once you have detected astigmatism you need to determine if it is coming from the objective or from the eyepiece or some other place. There is an easy test to see if the eyepiece is at fault. Line up on a bright star and look at the astigmatic image. WITHOUT ADJUSTING THE FOCUS, ROTATE THE EYEPIECE. If the astigmatism rotates with the eyepiece the astigmatism is in the eyepiece.

While you are rotating things to locate sources of astigmatism you should rotate the entire telescope. Don't just rotate the objective, because the fault is more likely to be in the attachment of the objective to the main tube than in the objective itself. If neither the objective nor the eyepiece seem at fault there may be two more places to look. If you have a star diagonal try checking the diagonal mirror. If it is not square to the optical axis the light from one side will be forced to travel a greater distance than the light from the other side.

There is one more part of the system to check. The observer's eye may be the problem. Just as the requirements for an astronomical telescope exceed those for other instruments the demands on the human eye are particularly high. You may have an astigmatic defect in your eye. If you wear glasses look at the prescription. If the boxes labeled "cylinder" have values the prescription is intended to correct for astigmatism. Try using your glasses as a test. That may fix the problem. Even if your eye doctor doesn't think that your astigmatism is serious enough to warrant correction it may show up under the demanding conditions imposed by a star image.

Before you make any decisions on tube length you need to determine the distance behind the objective lens and the eyepiece. If you simply put the eyepiece up against your eye as it would be if it were mounted in the finished telescope and hold the objective lens in its approximate position and look through the eyepiece you will be approximating the finished telescope. Fool around with the spacing until you can focus well on a distant object. The Moon would be ideal, but a distant hilltop will work. When you get a sharp focus you can have a second person measure the separation. Do this three times and average. If the difference between your measurements is more than 1/2 inch you need a better way because the average may not be close enough.

For a more accurate method attach, at right angles, a piece of white cardboard to the end of a stick. A yardstick or meter stick is fine. Slide the lens along the stick until you have as small and sharp an image of the Sun as you can get. Keep the lens square to the stick. Be careful not to set things on fire. Repeat the process three times and take the average. You may find it helpful to mask off some of the lens to reduce its effective diameter to reduce the heat and to reduce the brightness of the image so you can see it better. You can also use an image of the Moon.

However you got it you have the BFL of the objective lens. The eyepiece belongs positioned with its focus located at the focus of the objective. To locate the focus of an eyepiece put it against your eye as if you were using it. Stick your finger into the barrel of the eyepiece and determine where the end of your finger is when you can get a sharp image. Stick your finger in the eyepiece, not your eye. The end of your finger has located the focus. The image of the end of your finger will be much enlarged, so you may also have determined that your fingernails need cleaning.

The two focal points must coincide for an infinity focus. To achieve a closer (non infinity) focus the eyepiece will need to be moved back (away from the objective a small amount). The two focal points must also lie on the optical axis of your telescope. See the section on focusers for design ideas and a discussion of "throw."

By this point in the design and construction you should have decided on your focuser and whether or not you are going to use a star diagonal. Focusers and star diagonals are discussed in their own section. If you are using a star diagonal be sure to calculate its position to get the focus correct.

Most objective lenses (copyscope or otherwise) consist of at least two glass disks. One is made from "crown glass" and another from "flint glass." The crown component is positive. It will have at least one face distinctly convex. That is, it will operate as a weak magnifying glass. As an individual lens it will have a focal length of about 2/3–3/4 that of the two components combined. The other component (the flint) is negative; that is, it is a reducing lens. In combination the flint cancels part of the power of the crown but also has a color "error" that reverses – cancels – the color error of the crown. That is why you don't get the extreme color fringes that you got with the magnifying glass experimental telescope. The red end of the spectrum is folded onto the blue end, and the intermediate light ends up pretty much in the same place. All the light is combined, and we get white images when we are looking at a white object.

Most objective lenses are designed to have the positive (convex) face mounted outward – that is, pointing at the sky. If you do it the other way the flint intercepts the light first and disperses it somewhat as it goes into the crown. A 4-inch objective with both components at exactly 4 inches will, if mounted backwards, act as if it were about 3 7/8 inches in diameter. Some cases in which an objective lens is slightly under corrected for color can be fixed, or at least ameliorated, by reversing the positions of the two components.

Mounting a Copyscope

As a last planning step you need to decide how you propose to mount the telescope. This book has a section on mounts. In the case of a relatively small copyscope you may be able to use a camera tripod. Be sure that you have the telescope well balanced at the point where it will be attached to whatever mounting you chose to use. The standard for camera tripods is a 1/4 × 20 screw fitting. However you design the attachment make it solid. A piece of 1/8 inch or thicker metal with the 1/4 × 20 thread tapped into it, firmly attached to the body of the telescope, is minimal. Look at the description of the 80 mm refractor for another way of making the attachment.

The rings that you have seen in advertisements of commercial telescopes to attach either finders or mounting parts to the tube of a telescope can be approximated by automobile hose clamps. They come fairly large, and two can be connected to make one larger one. They can be shortened by simply putting a double fold in them. Just make two folds toward each other. Think it through so you get the doubled part on the inside of the curve. Use a little care while tightening. The clamps are designed to squeeze a hose full of hot liquid, under pressure, tightly enough to prevent leaks. It is not hard to crush things with them. You will see other applications of the lowly hose clamp in other chapters. They are truly useful things in telescope construction.

Chapter 5

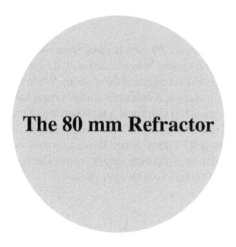

The 80 mm Refractor

A 3+ inch refractor can serve as a travel 'scope, a first 'scope project, or a practical 'scope to use at star parties. It is large enough to provide nice views but small and light enough to be easily transported and quickly set up. It can be built either as a simple, straight through 'scope or with a star diagonal or both.

Details for an easy tripod are included in this chapter. Both the telescope and the tripod can be put in an easy to make case to sit on the back seat of a car, inside a trunk, or even as checked luggage.

Telescope Description and Capabilities

This little telescope can serve as a "travel scope" with a lightweight tripod or as a finder scope for a larger telescope. With a 3-inch aperture it will pull-up most of the Messier objects at moderate power and can handle over 100 power for planetary and lunar observations. Most of the Lunar 100 (Charles Wood) objects are available. The main bands of Jupiter and the four Galilean moons can be seen with this little scope. The stripes and individual stars on a 3 foot by 5 foot American flag can be resolved (on a very clear day) at a distance of about 4 miles. Viewing anything through 4 miles of ordinary atmosphere has a lot in common with looking through a dirty aquarium. Because this telescope is one of the specific projects used by this book to provide examples of the use of the various components described in the book, it has been constructed in two versions: a straight-through version and a star diagonal version. Both versions and the tripod can be constructed with less than $75.00 spent on parts.

R.L. Clark, *Amateur Telescope Making in the Internet Age*, Patrick Moore's
Practical Astronomy Series, DOI 10.1007/978-1-4419-6415-1_5,
© Springer Science+Business Media, LLC 2011

Specifications and Components

Objective Lens

Focal Length 900 mm. Diameter 80 mm. It came from Surplus Shed at $29.

This is a good lens for its price. When tested with a Foucault device using auto-collimation it showed nice crisp images of the pinhole. When the complete telescope was tested against a star it showed reasonably uniform rings both in and out of focus.

The same source offers a 400 mm focal length version of the same lens. It would be suitable for a telescope of roughly 1/2 the overall length of this one. Operating at f5 it is close to being a RFT lens. Since the tube would be shorter it would be easier to mount as a finder on a medium length scope. The maximum power would be about half that of the 900 mm focal length version.

Objective Cell

A very important part of the design process for any refractor is the attachment of the objective lens to the main tube. Easy removal (for cleaning) is necessary, and the attachment must be very square. The combination of these two requirements produces a small complication. It would be relatively easy to get the lens attached square to the tube, but you need to be able to remove it and replace it without losing the alignment.

The lens comes in a nice cell that you might as well use, but the cell is not easily mounted in a way compatible with our two requirements. The easiest solution is to surround the cell with an outer cell that does satisfy our requirements. Since the main tube is to be a 3-inch, thick walled PVC an obvious solution is to make something out of a PVC fixture that will fit, squarely, over the 3-inch main tube. A 3-inch to 4-inch adapter cell is almost perfect. It fits well on the 3-inch tube and has enough space on the 4-inch end to accommodate the lens in the supplied cell. The PVC fixture is painted black and is on the right in Fig. 5.1. The lens in its black plastic cell, fits into the big end of the fixture with a little slop. The fixture has a slanted step where a square ring around the lens cell needs to sit. That allows the lens cell to tip around in the PVC fixture. It is very important to avoid this tipping.

A ring was cut off a piece of 4-inch PVC, about 3/8 inches of the 4-inch tube. It needs to fit into the larger end of the adapter fixture and slide down to the sloped step. It has a square edge to support the lens cell and is carefully leveled so it will serve as a good base for the objective cell.

Typical of PVC fittings the large end of the adapter has a slight taper. That is so that, when used for plumbing, a pipe pressed into the fitting will fit tightly. Because of that taper our ring has to be made into a split ring. Just cut about a 3/8 inch slot in the ring. That allows a little compression so it can be squeezed into position. It can be cemented in place with PVC cement. The supplied cell is slipped into the adapter to rest on your split ring and is held in place by two 1/4 × 20 machine screws taped

Fig. 5.1 Complete 80 mm telescope

into the sides of the PVC. They apply a little compression on the lens cell to keep it in place. In Fig. 5.1 the second of the two is hidden on the other side of the telescope. The additional screw shown, a little farther down the tube, secures the adapter to the main, 3-inch, tube.

The clear diameter of the front surface of the lens is just a little less than 79 mm. That works out to 3.1 inches. Are we compromising some of the area of the lens by attaching it to the end of a 3-inch tube? Yes, but even less than the 0.1 inch difference. The lens produces a converging cone of light in the tube. By the time that cone reaches the front of the 3-inch tube its diameter has been measurably reduced. Since the front of the tube will be at least 1 inch from the back of the lens the cone will be reduced by about 3/100 of an inch. If throwing away the remaining 7/100 of an inch of the lens seems to be a problem you can file away a little of the inside of the tube.

The importance of getting the objective lens square on the main tube can't be overstated. If the lens if set so that one edge is closer to the eyepiece than the opposite edge by 1/10 of an inch the image, back at the eyepiece focal plane, will be displaced by about 1 inch. That will have a very noticeable effect on the image as seen through the eyepiece.

The entire fixture (after it and the inside of the 3-inch PVC tube are painted dull black) can be slipped onto the main tube. There is a fabric called "flock" or "flocking fabric" that can be ordered from several online suppliers such as Edmunds Scientific and Anchoroptics. It can be glued to the inside of tubes to get the desired "non-reflective" result. The material comes in a self-adhesive variety that will allow you to avoid gluing. A company named Craftflocking sells the fibers and applicators so you can flock almost anything you like. Consider flocking your automobile. On second thought, forget that idea. Flocking tends to come in rather large quantities so combined orders are a good idea. Be sure you order black, as this stuff has other uses and comes in a variety of colors.

Main Tube

How long should the main tube be? Considering that it is easier to trim some off than it is to add onto the tube you should err on the long side. This little telescope will be made with two main tubes, since it is really being made as an example for this book. One main tube will be a "straight through" and the other will accommodate a star diagonal.

Since the "straight through" version avoids the diagonal mirror it can be expected to show a little less light loss and a slightly better image than the second version. The second version, with the star diagonal, will be significantly easier to use but involves the additional requirement of getting the first surface mirror, the diagonal, properly squared-up. If it is tilted, even a little, it will induce significant astigmatism. Optical problems such as astigmatism are much more obvious and bothersome in an instrument used on the heavens than in the same instrument used terrestrially. In that sense astronomical instruments need to be constructed with greater care than others. A star image that is more line than a nice, crisp, point is rather defective looking. It is a good idea to construct a "line-up" tool from cardboard to help getting the 45-degree angle and avoiding tilt in the diagonal mirror. The center of the focuser needs to be well centered on the diagonal mirror. The important point is to keep optical elements centered on the optical axis of the telescope and square to it. A cardboard tool (just a strip of cardboard that will fit down the tube, cut to a 45-degree angle) is useful in getting the diagonal correctly oriented. Such a tool is easy to make.

The nominal focal length of the objective lens is 900 mm. That converts to 35.43 inches and tells us that the focal point of the eyepiece needs to be 35.43 inches behind the objective lens. We have a problem – 35.43 inches behind what? Also, is the 900 mm really accurate?

The lens is only about 3/4 of an inch thick. It seems to be a simple two component design with a positive crown glass and a negative flint glass component. The two pieces of glass are either in contact with each other or, at least, very close. It is highly likely that the focal length is measurable from some point in the lens, probably about the middle. The 900 mm specification is almost sure to be pretty close. It is a mass production lens with a relatively short focal length. The longer the focal length of either a lens or mirror the more shallow the curves need to be on the surfaces.

The curves on the crown component of a lens like the one we are working with would have a much more easily measured radius than one with a much longer focal length. A focal length variation of 2 mm would be surprisingly large in our lens. We are more than safe if we place the focal point 35.5 inches behind the middle of the lens and allow 2 inches of adjustment for the focuser.

We have an additional inch or so of adjustment allowed by the fact that the PVC fixture used to contain the objective lens has about an inch of latitude in terms of how far it is pressed onto the main tube.

Eyepiece End and Focuser

The eyepiece end requires a little thought because we are building two versions. We would like to be able to do some switching around with the focuser to use it on both the straight through and star diagonal versions. We need a way of attaching a focuser to the 3-inch tube, again maintaining a square relationship while being able to take it apart.

A good solution is to make a plug with a hole for the focuser. The main part of the plug needs to fit into the 3-inch. tube and have a cap with diameter of 3 1/2 inches. There must be a hole through the whole thing to allow you to mount the focuser. The details of the hole depend on the details of the focuser. The assembled pair ready to be inserted into the tube is shown in 5.2. It fits the 3-inch. PVC main tube. The actual focuser came from Orion.

There are several options for the construction of the plug:

1. If you have access to a wood lathe it is very easy to turn the plug from almost any stable wood. Poplar is a good turning wood and is normally easy to obtain. You might have to do a "glue-up" job to get a large enough piece. The hole amounts to being a faceplate turning job.
2. You can cut several doughnuts from scrap plywood and glue them together.
3. Locate a PVC fixture fitting into or over the 3-inch tube and attach an end fixture. Drill the end fixture to accept the focuser.

However the plug is made it is important that it fit fairly tightly into the main tube. The plug can be held in place with a machine screw. Drill and tape the main tube for the screw. The screw does not have to extend into the plug; just exert a little pressure against it. That screw is visible just below the focuser in Fig. 5.1. The process of getting a 90-degree angle in the PVC tube is discussed in a later chapter.

Star Diagonal

Figure 5.1 shows the completed telescope in the star diagonal version. The straight-through version just substitutes a straight tube for the angled tube shown. As it is you could set it on a table and begin to try it. Without a mounting arrangement it is useless for the sky. Attempting to wave a 3-foot tube around in the sky just doesn't work. The travel tripod that is described a little later in this chapter will provide a good mounting.

The star diagonal needs a first surface mirror mounted at a 45-degree angle. Unlike the diagonal used in a Newtonian, this case (diagonal for a refractor) does not require that we pick a minimum size to reduce obstruction. As long as it is big enough to pick up the entire light cone in such a way that all light from the objective gets to the eyepiece field lens, you are OK. You can fill the entire tube if it is easier than cutting the mirror down. If you want the dimensions for the smallest diagonal providing full illumination you can modify the diagonal formula used for Newtonians as described below Fig. 5.2.

Fig. 5.2 Eyepiece end

Normally a diagonal is placed at 45 degrees to produce a 90-degree bend in the optical axis. A result of the 45-degree placement is the fact that the minimal diagonal would be shaped as an ellipse. The major axis (length) would be $1.414 \times$ the minor axis (width). Knowing that, you can do the arithmetic for the minor axis. The idea here is to make the entire objective visible from any point on the field lens of the eyepiece. You want to get all of the light coming through the objective relayed back to the eyepiece. The dimensions in Fig. 5.3 are:

L = Distance along the optical axis from the objective to the field lens of the eyepiece. For most eyepiece designs this is very close to the focal length of the objective.
D = Diameter of the objective.
d = Diameter of the field lens of the eyepiece.
l = Distance from the objective to the proposed diagonal.
X = Minor axis of the diagonal.

From similar triangles in Fig. 5.3 we have:

$$(D - d)/L = (D - X)/l$$

or

$$l (D - d)/L = D - X \quad X = D - l(D - d)/L$$

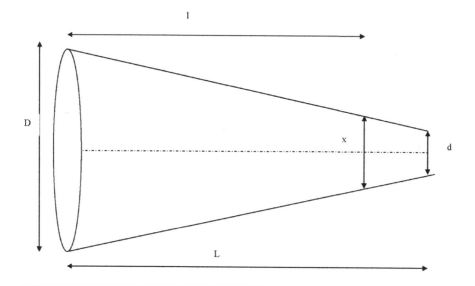

Fig. 5.3 Layout for refractor diagonal size

Finder

The figure shows a finder attached to the scope. It is just a reduced-size version of the main telescope. The objective lens is a projection lens, probably from a slide projector. The eyepiece lens is from a small projector. The focuser is constructed from natural gas fittings. It is less than 1 1/4 inch in diameter and will accept only the eyepiece for which it was designed. Focusing is done by turning the eyepiece on threads from the gas fitting. It has only about 1/2 inch of travel, but that is enough for such an extremely short focal length telescope.

The finder mount consists of a piece of oak about 9 inches long, about 1 1/2 inches wide, and about 3/8-inch thick. It is equipped with two thin rails glued to the bottom to keep it from rocking when held to the main tube by a hose clamp. On the surface of the 9-inch piece is another piece about the same thickness and width but only about 5 inches long. The screw at one end of that piece serves as a swivel and the screw at the other end can be tightened to position it. That provides the ability to adjust the finder right to left.

The up and down adjustment is provided by the square piece with a hole. It is glued to the swiveling piece. The hole accommodates a screw and wing nut going through a final piece that is attached to the finder tube.

We need to attach something to the telescope to serve as the intermediate between the telescope and the mounting on the tripod. Many commercial manufacturers use a flat plate arrangement with holes for machine screws or bolts. Others use a scheme with a key and slot. The key looks like an inverted keystone and is several inches long. The slot fits the key. A flat plate arrangement is the easiest for an amateur to construct.

A convenient arrangement is to equip the telescope with a flat plate with holes spaced in a rectangle 5 1/2 inches wide and 8 inches long. The holes should be 3/8 of an inch diameter or even 1/2 of an inch. The actual mounting surface is equipped with 1/4 × 20 studs sticking up from the surface.

The plate permanently attached to the telescope should be placed as tight against the telescope tube as possible. It needs either to be located at the balance point of the telescope, or, if the balance point may move, the plate needs to be moveable. Changing from a straight-through design to a star diagonal will move the balance point. Changing eyepieces can change the balance point. However the plate is attached it is very important that it be solidly attached. Neither it nor the tube should be allowed to bend or flex. The 5.5 by 8-inch "footprint" is none too big.

An alternative mounting attachment scheme takes an idea from a common way of handling the mounting of Dobsonian telescopes. The idea is to attach a pair of stout dowels to the sides of the tube to stick out a couple of inches. These are placed at the center of gravity of the telescope so no counterweights are required. On a Dobsonian the center of gravity is fairly near the bottom of the tube, so this arrangement often puts the eyepiece at a reasonably convenient level.

Both the Dobsonian design and our adaptation for refractors work most simply in an alt-azimuth mode. They can be adapted to equatorial mode, and, indeed, for Dobsonians, there are platforms that, by virtue of software, operate as if the mount was equatorial.

The arrangement in Fig. 5.1 shows both approaches combined into one. The plate is the bottom of a cradle attached to the telescope tube by a hose clamp. On the side of the cradle you can see one of the dowels where it rests in a groove in the tripod head. The combined design requires that the plate be a little further from main tube than it would be if it were the only mounting arrangement provided. The chapter on the 4-inch copy scope refractor has an illustration of another cradle.

Tripods are popular ways of mounting telescopes. However, they tend to be expensive, heavy, flimsy, or some combination of the three. Various homemade designs tend to have problems with folding the legs so the tripod can be moved. An effective way to get past that problem is to make the legs removable.

Here is an easy travel tripod that will cost you almost nothing to make. It weighs very little and is much steadier than the normal, aluminum-legged camera tripod. It will take only a couple of hours to put together, and, in addition, it gives you a productive use for the little pieces of scrap plywood you have been keeping. The couple of hours of assembly time is punctuated by an overnight glue-setting spell.

Here are the materials you will need:

- Scrap plywood any thickness is fine. A mixture of thicknesses is OK. This stuff will become the tripod head.
- Three 2 × 2's or two 2 × 4's long enough for the legs. 40 inches is about right. It depends on how you intend to use this thing.
- Three hanger bolts, which are the same thing as toilet hold-down screws. These are about 3 1/2 inches long with a screw thread on one end and a machine screw thread on the other end. They will hold the legs to the head.
- Wing nuts and large washers to fit the hanger bolts.

Lay out an equilateral triangle on some reasonably strong paper. Old file folders work fine. About 8 inches on each side is reasonable, but suit yourself. Mark the midpoint of each side and measure out from it 3/4 inches on each side. We are assuming that you will make the legs 1 1/2 inches square. From the marks you just made go out another 1 1/2 inches in each direction. Next draw three lines cutting off the corners of your triangle (Fig. 5.4).

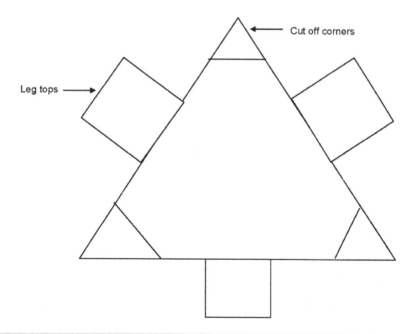

Fig. 5.4 How to cut the corners and tops of the legs

Cut off the corners and trace it onto the plywood. Set the tracing back from the edge an inch or so. Cut the plywood to shape, flaring out 30 degrees so the bottom is larger than the top. If you are using 3/4 inch plywood the bottom of the plywood will be larger than the top in all directions by 0.43 inches. These are easy cuts to make on a table saw or by hand.

If you are using a saber saw just set the base to 30 degrees and cut, following the marks on the top. The bottom will take care of itself.

If you are using a table saw just set the table to 30 degrees and follow the marked lines.

You now have the top piece of a sandwich. Set it on some more plywood and trace it. You will get the same shape but larger. Cut again and continue. How many layers you need depends on the thickness of the scrap plywood that you are using. You need to make it at least 4 inches thick. Note that, except for the top layer, you can make the layers out of smaller pieces if your scrap plywood is in small pieces.

Line up the sandwich and glue and clamp it together. You now have a truncated pyramid. The glue may act as a lubricant, letting the layers slide all over each other if you attempt to glue many layers at once. You may get a neater job if you glue it up in pairs. Unless the lineup is very good you may get irregular sides on your pyramid. Use 50 weight sandpaper on a disk sander or a wood rasp to get it evened out (Fig. 5.5).

If you have access to a band saw you can be less exact with the glue job because you will be able to slice the thing into proper shape. Remember, meat cutters use band saws to make chops, etc. Be careful!

Fig. 5.5 A completed truncated pyramid

Make the legs by ripping the 2 × 4 down to two pieces 1 1/2 × 1 1/2, or trim the 2 × 2's to 1 1/2 by 1 1/2. The length is entirely up to you. You can use a decent hardwood such as American red oak if you like. Just make sure, particularly if you are using pine, that there are no significant knots or flaws in the legs.

From the leftovers resulting from making the legs cut six pieces long enough to fit the slope of your truncated pyramid. These need to be screwed and glued to the sides with just enough space between them to accommodate a leg. Take a piece the same size as a leg and lay it against the side of the truncated pyramid, centered on the side. Trace the sides. The pencil marks will give you something to line the short pieces up against. If the slots between the short pieces are a little too narrow that is OK. You can trim the legs to fit.

Once the glue has set trim the pieces just added so that the tops and bottoms are flush with the top and bottom of the pyramid. See Fig. 5.1 shows the pyramid with the short pieces attached and trimmed. Next put a leg into its slot, sticking just a little above the top of the pyramid, and drill through the leg and just a little into the pyramid for the hanger screws. Deepen the holes in the pyramid with a fine drill to make a pilot hole. You are drilling both at the same time to make sure that the holes line up. It doesn't matter if the three legs and their attachments are slightly different. You only need to mark the legs and their slots.

If you want a field tripod and one to set on a table make a second set of shorter legs. In the process of making the pyramid it is likely that the flat spots between the side pieces can turn out to be slightly convex. Since we are using a single screw to secure each leg convexity can result in rocking of the legs. It is better to scoop out a little wood to get them concave than to leave them convex. After you get the legs fitting nicely into their slots it is likely that they will be slightly different sizes. Number the legs and the slots to make assembly easy.

Assemble the whole thing and trim the tops of the legs to match the top of the pyramid and the bottoms of the legs to fit the ground. Sand it and apply whatever finish you like.

This tripod is going to carry a relatively light load, so a "lazy Susan" is an effective swivel. Use as good a one as you can get and as large as will fit of the top of your truncated pyramid. The arrangement shown in Fig. 5.1 uses a flat plate (plywood) screwed to the top of the lazy Susan. Attached to it are two diagonal pieces and a pair of 2 × 2's to support the telescope. The 2 by 2's are drilled to take the dowels on the cradle. Both 2 × 2's are sawed down the middle to allow the dowels to be placed in the resulting half holes. On one side (the side showing in Fig. 5.1) the cut-off piece is hinged and equipped with a screw to allow a little friction to be applied to the dowel joint. On the other side the half hole is left open and the dowel, on that side, simply rests in the half hole.

The 4-inch refractor chapter (Chap. 7) illustrates a heftier version of the same top design. The lazy Susan used in the lightweight tripod is replaced by a pipe joint, that can be angled to provide an equatorial axis and the whole thing is on a pipe filled with concrete.

Chapter 6

Newtonian Telescopes

Newtonian reflectors are still the most popular form of amateur telescope, whether it is a homemade telescope or a commercially made instrument. This chapter provides general guidance on the design and construction of Newtonian telescopes using web-sourced parts.

Typical Newtonian Telescopes

A 6- or 8-inch-diameter Newtonian telescope was, in the days when the builder had to grind and polish the objective mirror, almost the "standard first serious telescope." Today, the availability of appropriate mirrors from a variety of Internet sources makes the hand grinding and polishing unnecessary. In fact, this is one of the greatest advantages that a modern amateur telescope maker has – the availability of good Newtonian mirrors at very reasonable costs. A reasonable estimate for grinding a 6-inch mirror is 12–16 h. The polishing can go to as much as 40 h. An 8-inch mirror is probably 1 1/2–3 times as much work. In both cases you will need to send the mirror for coating (aluminizing). That will cost $30–$50 plus shipping. You can purchase a reasonable 6-inch f5 parabola for about $50 or an average 8-inch one for about $75. See the section on quality and testing for hints on how to purchase such a mirror.

Another change is based on convenience. The typical amateur-ground and polished mirror was an f6–f8 mirror. The modern trend with commercial telescopes has become "faster." Modern purchased mirrors are often f5 or faster. The difference between an f8 mirror and a sphere is very little. One could get good performance out of an f8 sphere without worrying about the mysteries of "parabolizing" at all. The faster mirrors, some as fast as f3.5, demand that the parabolizing task be done and

R.L. Clark, *Amateur Telescope Making in the Internet Age*, Patrick Moore's
Practical Astronomy Series, DOI 10.1007/978-1-4419-6415-1_6,
© Springer Science+Business Media, LLC 2011

done well. The principle advantage of these fast mirrors comes from their portability and convenience in use. An f8, 8-inch mirror has a focal length of 64 inches. That requires a tube length of about 60 inches. If the telescope is pointing straight up and the bottom end is near the ground the eyepiece will be more than 5 feet off the ground. That is out of reach for many people. A 12-inch f8 requires the use of a ladder and can hardly be regarded as portable.

Another change is based on weight. The advent of the Dobsonian design has led to serious – and successful – efforts to have relatively large Newtonian telescopes be fairly portable. Since the mirror is the heaviest single part of a Newtonian telescope the telescope can be made considerably lighter by using a thin mirror. The older mirrors tended to be 1/4–1/5 as thick as their diameters. An older 12-inch mirror would be as much as 2 1/2 or 3 inches thick. That makes a 0.164 cubic foot hunk of glass that may weigh as much as 30 lbs. Add to that the rest of a telescope and you are stretching the limits of "portable." A later chapter of this book describes how to make a large Newtonian. The mirror selected was a 12.5-inch f6 available from eBay. In that example the mirror is rather thick by modern standards and portability is not a serious consideration.

Following are some ideas on building various Newtonian reflectors mostly from web parts. Two telescopes are used for illustration. One is a 6-inch f5 instrument built around a mirror from eBay (probably from China). The other is a 6-inch f9 instrument built around a handmade mirror. There is an entire chapter devoted to a telescope built around a 12.5 inch f6 mirror from eBay.

Figure 6.1 shows a 6-inch f5 Newtonian with worldwide web-sourced optics. The main mirror, the focuser, the diagonal, and the parts for the finder all came from

Fig. 6.1 A web-based short Newtonian with a wood tube

various web sources. The tube is homemade from wood. Details on tube making can be found in a later chapter.

The Objective Mirror

The mirror is the heart of your telescope and is the part that you are least likely to ever replace. When you want a bigger or better telescope you will probably build another. In the days when all amateurs ground and polished their own mirrors the standard first scope was a 6-inch Newtonian with an occasional 8-inch mirror done by the most ambitious. The most common focal ratios were f6, f7, f8, or even f9. The reason for the relatively long focal lengths of 36, 42, 48, or 54 inches producing large f numbers was the fact that parabolizing is either fairly easy or not needed at all for relatively long focal lengths.

If you are reading this book you most likely intend to purchase your mirror. For that reason it is suggested that you consider an 8-inch, or even 10-inch, mirror. As far as f ratio is concerned you have a slightly more complicated choice. You need to balance the following conditions:

- The most demanding step in making a mirror is the parabolizing. This is true if you are doing it yourself or if you are purchasing the mirror. It follows that, if quality is sacrificed anywhere in the process, it is likely to be there.
- Your tube length will be approximately the same as the focal length. Do you want to get this thing into and out of the back seat of a VW?
- Mirrors of the same quality are probably going to be more expensive the bigger they are and the shorter the focal length.
- The bigger the telescope the heavier it is going to be.
- The thinner the mirror the less it will weigh but the more critical the evenness of the support offered by the cell will be.
- Longer focal lengths and larger mirrors will allow the use of a diagonal that provides less central obstruction.

All of these conditions may come together to indicate an 8-inch f6 or a 10-inch f6. You might go to f5 for each, but expect to pay a bit more to have a decent chance of ending up with a good mirror. If you can find one and are willing to deal with the extra length an 8-inch f7 would be a good deal. The choice of a mirror is, essentially, a personal one.

How big a secondary (diagonal) do you need? If it is too small you will be throwing away part of the light coming from the primary. Why build an 8-inch instrument if all you are going to use is the inner 6 inches?

If it is bigger than needed you have introduced more central obstruction than is needed. You will have to tolerate at least 2% obstruction.

If the secondary is to be placed at 45 degrees to the axis of the primary mirror (this is the standard orientation) it should be elliptical, with the major axis 1.41+

times the minor axis. The 45 degree condition is, by far, the most common choice but is, certainly not the only possibility. Pythagoras tells us that the long side of a right triangle with 45-degree corners is the square root of 2 times the length of a side. The tangent function of 45 degrees is 1.41+ as a result of the same geometry. If you need some other angle the tangent function is easy to look up. The shape is always an ellipse because viewed at the appropriate angle an ellipse looks like a circle. What will change is the relationship between the major axis and minor axis.

Diagonal dimensions are pretty easy geometry, but that doesn't keep them from being commonly done wrong. To see the correct geometry first imagine or draw the situation in which the diagonal is not present.

Figure 6.2 represents a common incorrect perception:

The problem can be seen by imagining yourself positioned at the vertex of the triangle so that your eye replaces the eyepiece. You would see the entire primary, so your eye will receive all of the light received by the primary. If you move either up or down, some of the primary will be hidden. The result of this geometry is that only the center of the field of view is fully illuminated.

A formula giving r, half the minor axis size of a diagonal that allows full illumination of only the center of the image is:

$$r = R \times f/F$$

where R is the radius of the objective, F is the focal length of the objective, and f is the distance from the focal point of the objective to the diagonal. That formula yields a value that is too small. The only reason for presenting it is the fact that it represents the most common error. If, for example, you had an f5, 6-inch telescope with the value of f at 5 inches you would get:

$$r = 3 \times 5/30 = 0.5$$

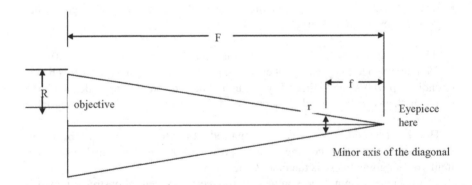

Fig. 6.2 Diagonal to small

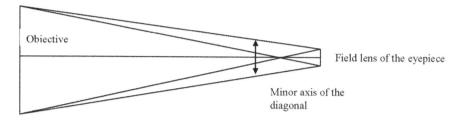

Fig. 6.3 Diagonal size, better geometry

and your diagonal would be only 1.0 inches wide. Since it looks like a circle from the point of view of the eyepiece its area would be only about 0.75 square inches, resulting in a 2.7% obstruction.

Figure 6.3 represents a more correct geometry. From any point on the eye lens of the eyepiece the entire objective is visible. Thus the entire eye lens is fully illuminated.

Now we need to decide how large a part of the eye lens we want to fully illuminate. For traditional eyepieces 1/2–5/8 inch is probably plenty, but some of the more modern (expensive) designs can need more. Look at and measure the eye lens of the fattest eyepiece that you are likely to want to use.

If you intend to do prime focus photography you need to think of the eye lens as the size of your film or your receptor. If you intend to do digital photography or large format photography you may even decide that the size required for complete illumination means that a standard Newtonian design is not acceptable. A refractor or a catadioptic design may serve better in that you won't have a big central obstruction.

The following table applies to standard Newtonians with diameters (Obj Dia) starting at 6 inches and going to 12 inches in 2-inch steps.

Obj dia	Obj Fl	Dia FEF	BackD	Min Ax	Area Obj	Area Blkd	Obs (%)
6	30	0.5	4	1.233	28.26	1.19	4.23
6	30	0.5	5	1.417	28.26	1.58	5.57
6	30	0.5	6	1.600	28.26	2.01	7.11
6	36	0.5	4	1.111	28.26	0.97	3.43
6	36	0.5	5	1.264	28.26	1.25	4.44
6	36	0.5	6	1.417	28.26	1.58	5.57
6	42	0.5	4	1.024	28.26	0.82	2.91
6	42	0.5	5	1.155	28.26	1.05	3.70
6	42	0.5	6	1.286	28.26	1.30	4.59
8	40	0.5	5	1.438	50.24	1.62	3.23

				(continued)			
Obj dia	Obj Fl	Dia FEF	BackD	Min Ax	Area Obj	Area Blkd	Obs (%)
8	40	0.5	6	1.625	50.24	2.07	4.13
8	40	0.5	7	1.813	50.24	2.58	5.13
8	48	0.5	5	1.281	50.24	1.29	2.57
8	48	0.5	6	1.438	50.24	1.62	3.23
8	48	0.5	7	1.594	50.24	1.99	3.97
8	56	0.5	5	1.170	50.24	1.07	2.14
8	56	0.5	6	1.304	50.24	1.33	2.66
8	56	0.5	7	1.438	50.24	1.62	3.23
8	64	0.5	5	1.086	50.24	0.93	1.84
8	64	0.5	6	1.203	50.24	1.14	2.26
8	64	0.5	7	1.320	50.24	1.37	2.72
8	73	0.5	5	1.014	50.24	0.81	1.61
8	73	0.5	6	1.116	50.24	0.98	1.95
10	50	0.5	6	1.640	78.5	2.11	2.69
10	50	0.5	7	1.830	78.5	2.63	3.35
10	50	0.5	8	2.020	78.5	3.20	4.08
10	60	0.5	6	1.450	78.5	1.65	2.10
10	60	0.5	7	1.608	78.5	2.03	2.59
10	60	0.5	8	1.767	78.5	2.45	3.12
10	70	0.5	6	1.314	78.5	1.36	1.73
10	70	0.5	7	1.450	78.5	1.65	2.10
10	70	0.5	8	1.586	78.5	1.97	2.51
10	80	0.5	6	1.213	78.5	1.15	1.47
10	80	0.5	7	1.331	78.5	1.39	1.77
10	80	0.5	8	1.450	78.5	1.65	2.10
10	90	0.5	6	1.133	78.5	1.01	1.28
10	90	0.5	7	1.239	78.5	1.20	1.53
10	90	0.5	8	1.344	78.5	1.42	1.81
12	48	0.5	7	2.177	113.04	3.72	3.29
12	48	0.5	8	2.417	113.04	4.58	4.06
12	48	0.5	9	2.656	113.04	5.54	4.90
12	60	0.5	7	1.842	113.04	2.66	2.36
12	60	0.5	8	2.033	113.04	3.25	2.87
12	60	0.5	9	2.225	113.04	3.89	3.44
12	72	0.5	7	1.618	113.04	2.06	1.82
12	72	0.5	8	1.778	113.04	2.48	2.19
12	72	0.5	9	1.938	113.04	2.95	2.61
12	84	0.5	7	1.458	113.04	1.67	1.48
12	84	0.5	8	1.595	113.04	2.00	1.77
12	84	0.5	9	1.732	113.04	2.36	2.08
12	96	0.5	7	1.339	113.04	1.41	1.24
12	96	0.5	8	1.458	113.04	1.67	1.48
12	96	0.5	9	1.578	113.04	1.96	1.73

The f ratios go from f5 to f9, so the focal lengths (Obj Fl) run from 5 times the diameter to 9 times the diameter except in the case of 12-inch objectives. For those the range is 4–8 times the diameter.

The column headed BackD measures the distance from the diagonal to the image plane. Most eyepieces require that the image plane be just slightly outside of the field lens. You can get a good approximation for any particular eyepiece by holding it to your eye and sticking a finger down the barrel. At some point the end of your finger and the dirt under the nail will come into focus, normally within an inch of the field lens. That position is where the objective's image plane must be placed. BackD starts at 1 inch greater than 1/2 the diameter of the objective, goes to 2 inches greater, then 3 inches greater. BackD is a rather critical measurement. It has to be at greater than 1/2 the tube width to get the image to the outside. We arbitrarily picked 1 inch greater so that the image falls 1 inch outside the tube if the tube is exactly the size of the objective mirror and has no wall thickness. That is clearly a minimum condition.

Min Ax is the most important result. It is the minor axis of an elliptical diagonal that will provide a 1/2 inch diameter circle of full illumination. This will fill a field lens with a diameter of 1/2 inch. An absolute, theoretical, lower limit would provide full illumination only at the very center. If you adjust the table values by subtracting 0.5 inches you will be at that impractical limit, but you might get away with subtracting 0.2 inches if you intend to use only high power eyepieces with small (0.3 inches) field lenses. It should be obvious that primary focal plane photography with a small, fast Newtonian will require very a big obstruction.

Area Obj is simply the area of the objective mirror.

Area Blkd is the area blocked by the diagonal. The diagonal is assumed to have a major axis $1.414 \times$ the minor axis and to be mounted at 45 degrees so it presents a circle to the light cone.

Percentage of Obs is the percentage of the objective mirror that is occluded by the secondary (the diagonal) for an object in the middle of the field of view.

From the geometry and the formula you can see that getting the eyepiece as close to the diagonal as you can will pay off in reducing the size of the required diagonal and, thus, reducing the size of the central obstruction that it becomes. This is an important consideration. A 5% obstruction will have very little effect on the image. It will be slightly "stirred-up" in the center, but most of the time it won't be noticed. If you are trying to resolve one of the little craters on the floor of Plato you may, instinctively, move the image a little off the center. If you allow a 7% obstruction you will notice it.

So-called "low profile" focusers are designed to get the eyepiece as far "in" as possible. Be careful not to allow any part of the focuser to hang into the tube. It won't impinge on the image directly, but it will serve to shade part of the mirror and reduce its effective area.

If you will examine the table carefully you will see that minimizing BackD is most important for the smaller objectives and the shorter focal lengths. The most extreme case in the table is the f5 6-inch telescope (diameter 6 inches, focal length 30 inches), yet that specification is probably the most common one offered on eBay. This makes for a very handy, reasonably short, telescope. From that observation we can see that "low profile" focusers are most valuable with fast (low focal ratio) telescopes because they allow the eyepiece focal point to "get in" close to the center of the tube.

One of the problems associated with some wooden tubes as described elsewhere in this book is the fact that, because of the supporting rings inside, you need a tube somewhat larger than your mirror. That pushes the focuser and eyepiece out, away from the optical axis. You should consider building a flat spot on the side of the tube to accommodate the focuser. The same problem may occur with PVC tubes. A 6-inch tube is not large enough for a 6-inch mirror with its cell. The next available size from your local hardware store may be 8 inches. From the outside of a 8-inch PVC tube to the optical axis (center of a mirror) is 4 1/4 inches. Without a flat spot you will be pushing 6 inches for a BackD. A flat spot will allow you to get rid of almost 1 inch of distance to the optical axis.

Mirror Mounting

The objective mirror for a Newtonian, a Cass, a Gregorian, or any variation needs to be evenly supported. Glass bends! The support mechanism is normally called "a cell." This requirement and terminology also applies to refractor lenses. As a matter of fact the bending that occurs due to gravity acting on the glass when the telescope is pointing in various directions serves to limit the reasonable size of refractors to about 40 inches. Beyond that size the weight of the lens causes it to sag so much as to seriously mess up its performance. In a refractor all of the support must be at the edges. In the case of a reflector we can support the main mirror from the back. Since we can deal with back support we have a lot more options. We want to provide uniform support without any twists, bumps, or bends. Except in the case of quite small mirrors don't glue it down. Even in small cases such as a diagonal for a small Newtonian there is a real danger that temperature changes will affect the glass and the surface it is glued to differently. The result will be a warped mirror and distortion. If you try to glue very lightly or with a flexible glue such as a bathroom adhesive the mirror is likely to fall off its backing.

Anything that falls apart in a telescope is sure to land on the objective mirror and break it. It is much better to use some sort of non-pressure type clips. These will obstruct some very tiny part of the mirror, but you will have a secure arrangement. The clips made for hanging ordinary mirrors on a wall will work very well if you have space for them. Generally, in the case of an objective mirror, you will have enough space alongside of the mirror. You will probably need to put some washers under the clips to make them high enough to fit your mirror. Put something compressible under the mirror. For small mirrors felt or an old computer mouse pad works very well.

In cases where you can't use the mirror clips consider the following:

Make a little hardwood cylinder with a diameter the same as the minor axis of your secondary mirror. Cut the cylinder off at a 45 degree angle. The cut face will

Fig. 6.4 Diagonal mirror held by clips to a wood backing and a single vein spider

now be an ellipse to match your mirror. Use any sheet metal (a coffee can lid will work) to make clips that attach to the side of your cylinder and wrap up over the edge of the mirror. Allow space for the felt, mouse pad material, whatever. The fit should be almost loose, a minimum of pressure anywhere. In Fig. 6.4 the clips are the black pieces of sheet metal.

A Spider in Your Telescope

The secondary mirror needs to be, somehow, suspended in the middle of the main tube to direct the primary image out the side, where it can be examined with an eyepiece. It would be nice and would quiet a lot of loud dissension if we could just suspend it there. Unfortunately such magic does not work. Something must hold it where we want it. That something is normally called a spider. If you do a web search you will find multiple designs, each with its advocates. The diagonal mirror, once attached to something to hold it, must be located inside the telescope tube in a way that allows adjustment but provides a solid enough positioning as to minimize the adjustments. You don't want to have it knocked out of alignment by simply moving or bumping the telescope. But, when it is out of alignment, you need to be able to correct it fairly easily. There are a few things that most knowledgeable folks will agree on:

- The design of a spider has an effect on the image. For example the 3 or sometimes 4 spikes that you have seen sticking out from star images come from the spider. Real stars don't look that way.
- The thicker the supports the worse the spikes.
- If you ignore them they won't go away. However, you can learn to tolerate them, like a loud kid in a restaurant.
- You get as many spikes as you have supports. (Actually, you can get twice as many if you get the supports out of line.)
- If the supports are curved you won't see the spikes, but the light that would have go into the spikes gets scattered around and reduces the contrast of the entire image. See Fig. 6.5.

It may seem a bit odd that a pretty good-sized central obstruction has very little effect on the image, while some very thin, straight obstructions have such an obvious effect. Most telescope makers simply accept the defect, but there are several things you can do about it:

- You can use a single vane rather than 3 or 4. Such single vanes can be purchased or easily built. See Fig. 6.4. It shows a single-vane spider with the secondary mirror attached to a hardwood prism with metal clips.
- The stalk is made from two pieces of sheet aluminum. One piece has a right angle bend for attachment to the inside of the tube and, at the other end, a slot

Fig. 6.5 Three-vane spider (Courtesy of Gary Frishkorn, Westminster Astronomical Society)

to accept a machine screw attaching it to the other piece of sheet aluminum. The second piece has a hole and machine screw with a nut at one end and a right-angle bend with holes for screws attaching it to the wood prism to which the secondary mirror is attached. Figure 6.5 shows the spider disassembled.

- You can use curved vanes. They will almost do away with the diffraction spikes, but they are harder to build and tend to reduce contrast.
- You can use multiple curved vanes. A very elegant idea is to make curved vanes using strips cut from the ends of sheet metal stovepipe. See Fig. 6.6. It shows a three-vein spider with curved legs.

Be sure you have accommodated all possible degrees of motion into your design. You need to be able to adjust it in and out of the tube, that is, toward and away from the primary mirror. That adjustment is necessary to get the diagonal "under" the hole in the tube for the focuser. Once the diagonal is positioned in the plane perpendicular to the optical axis and lined up with the focuser hole, you need to be able to shift it back and forth in that plane to get it directly under the hole. Then you need to be able to adjust the "tilt" both left and right and up and down so you are reflecting the light cone squarely out the focuser hole.

All these adjustments constitute most of the process called "collimation." It is suggested that you read the section of this chapter devoted to collimation before you make final decisions regarding your spider, so that you can have a clear mental picture of the ways you need to be able to push the diagonal around.

Fig. 6.6 Parts for a single vane spider

Spiders, diagonal holders, and entire assemblies including the diagonal mirror are available from several Internet sources.

Mirror Cells

Attaching an objective mirror at the bottom of the tube is similar in many ways to mounting the diagonal but with four important differences:

- The objective mirror is a lot bigger and heavier than the diagonal.
- The positioning has to be more precisely adjustable.
- The positioning adjustments must be made from outside of the tube.
- Any distortion of the primary will have a very serious effect on the quality of the image.

The positioning precision and accessibility from outside the tube can be accomplished with the "collimation screws." These are located on the outside of the bottom of the tube and are used to adjust the exact tilt (or lack of tilt) of the mirror. They are normally positioned at the points of an equilateral triangle. One way to build a fairly simple rig that works well for Newtonian mirrors up to about 10 inches is to attach the mirror to a piece of plywood (well padded) that is itself attached to a piece of thick sheet metal drilled and threaded for the collimation screws. See Fig. 6.7. The mirror is sitting on an old computer mouse pad and held in place on the plywood base by sheet steel clips that exert almost no pressure on the mirror. A 28-gauge soft steel sheet from your local hardware store does just fine. The sheet metal triangle is simply screwed to the back of the plywood base with something between it and the base to hold it off the plywood by 3/16–1/4 inches. It could be a piece of Masonite^{TM} or just a few washers on the wood screws used to hold it in place.

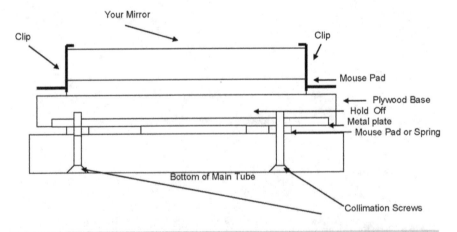

Fig. 6.7 Mirror cell

The sheet metal should be drilled and tapped for the collimation screws. Make sure that the metal is thick enough to hold the threads. A 1/10 or even 1/16 piece of aluminum is fine. Be sure that the holes line up with the holes in the bottom plate of the telescope. An easy way is to clamp the sheet metal to the bottom piece of the telescope and drill through the pair at once. 1/4 × 20 machine screws will work well. Don't forget that the drilled holes in the metal need to be a bit smaller than the bottom holes to accommodate the threads.

Countersink the holes in the bottom deeply so that the screw heads will fit well below the surface. When the telescope is handled to mount it, etc., the bottom will get bumped. If the screw heads get bumped that will move the mirror, so protect them by giving each its own little foxhole.

Between the bottom plate and the metal place small compression springs or washers made from more of the mouse pad. They will keep the whole deal under enough tension to pull the screw heads down into their countersinks and give you a positive adjustment.

The attachment of an objective mirror to the bottom end of the tube of a Newtonian telescope can also be accomplished using some Z-shaped fixtures instead of an aluminium sheet. The fixtures are screwed to the plywood backing of the mirror and positioned with machine screws in the same way as the aluminum sheet in the above example. Both ends of the clips have a hole drilled in them. One end is tapped to fit the machine screws. The other end is just a hole for a flat head screw that goes into the plywood mirror backing. Figure 6.8 shows three clips; one has the tap still in the hole. The other two have already been tapped.

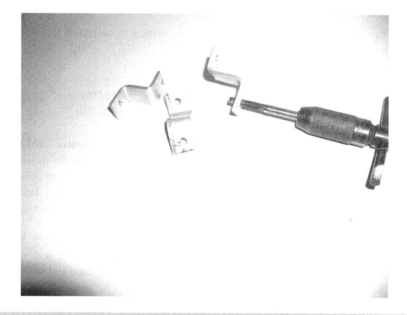

Fig. 6.8 Z collimation clips for a main mirror

The aluminum strap material used for the illustrated fixtures is about 1/8 of an inch thick and 3/4 of an inch wide. It is available online from Small Parts, Inc. It takes a little planning to get usable Z shapes out of this stuff. Soft steel strap material with about the same dimensions can also be used but takes the same planning. You need something like a machinest's vice (that has jaws at right angles to the grip). You will put the work into the jaws of the vice and bend it to 90 degrees with a hammer. Keep in mind that you need pretty solid vice jaws and a well defined right angle.

Here's a suggested procedure:

1. First mark the material where the bends should go and where the holes should go.
2. Drill all the holes and tap the appropriate holes. Then saw the pieces apart.
3. Next make the bends. Be careful to get the bends as square as you can and don't bend any closer than about 3/16 of an inch from a hole. Holes may still be distorted in the bending process, so renew the threads after the bends have been made. You can use a block of hardwood in the vise with the work piece to avoid distorting the first bend while making the second bend.

For mirrors above 10 inches you are best advised to buy a cell. They often show up on eBay, AstroMart, etc. Often mirrors cells are sold on eBay and the like with their mirrors. That gives you a cheap cell, and you are guaranteed that the cell and the mirror match. You may find mirror, cell, secondary holder spider, and secondary mirror available as a single item. Those deals can be good. Often someone has assembled the stuff for a telescope and changed their mind. Check that the secondary mirror is somewhere near to the correct size. There is no point in buying someone else's mistake. See Chap. 9 for an illustration of a typical mirror cell attached to a baseboard.

The bigger and the thinner the mirror the more critical having a cell that provides even support becomes.

Simple Collimation

Collimation is the process of getting the optical components properly lined up. It is an activity that must be carried out on a finished telescope, but the degrees of freedom and adjustment must be built in to the telescope if you hope to collimate it when it is finished. The process is outlined here because you need to check the feasibility of collimation as you build. Make sure that all of the adjustments work prior to finishing.

Collimation is critical. It is particularly critical with short focal length Newtonians. If an instrument is even slightly out of collimation you will not be able to focus a star to a minimum point. Spreading the light of a faint star will make it invisible. Poor collimation can cost you a magnitude.

Here is a procedure that works well.It can be refined, but this will get you to the point of decent performance.

1. Get the mirror square to the tube. You can waste a lot of time if you don't start out with a squared off mirror. This is likely to happen with a home-built cell or home-built attachment of the cell to the tube. (It is easy to get as much as 1/8 of an inch tilt in the mirror. No matter what you do the result will be GROSS astigmatism.)

   An easy way to roughly check is to remove the diagonal and its spider. Stand off about 15–20 feet from the front end of the tube. If you look square down the tube (so you see equal amounts of the inside of the tube) you should see a reflection of your eye in the center of the mirror. At 20 feet you will be pretty good if you get the reflection of your head (wrong side up) in the center. As this point the mirror is pretty close to square. The last step in collimation will refine the "squareness."

2. Now install the diagonal with its spider. Get the diagonal in the center of the tube. Mark the center of the back of the diagonal holder. The mark should correspond to the center of the diagonal mirror. Make a disk from file folder stock to just fit inside the tube. Make a small hole in the center of the disk. Put it in the tube. The hole should line up with the mark on the back of the diagonal holder.

3. Get the diagonal positioned so that when you look down the center of the focuser tube you see it in the center of the focuser tube. Move it up and down the tube until it is centered.

4. Now you can do the fairly standard procedure of getting a centered image of your eye through the focuser tube. When you get a centered image of your eye (viewing at the center of the focuser tube) you see the entire mirror, centered in the focuser tube. Within that you see an image of the focuser tube, within that you see an image of your eye. When everything is centered, you are pretty close.

Finding the center of the focuser tube can be a problem. If you have a broken eyepiece you can replace the lenses with a plug that has a small hole (about 2 mm) in the center. The black plastic cans used for 35 mm film work. Cut out a hole in the bottom almost as large as the bottom and make a hole in the cap. If you push this elegant rig into a 11/4 inch focuser it will center itself and force you to be looking through the center.

Next try the complete telescope on something like a point image. A star is fine, but messing around with fine adjustments in the dark is awkward. A point reflection of the Sun (from a Christmas ornament or telephone insulator) will do fine or something like a fleck of white on the side of a telephone pole about 100 yards away. If you still have a little astigmatism you will see a pretty good image at best focus, but, at just a little inside and outside of focus the point will elongate like this:

I

or this: ▬

The direction of the elongation *outside* of best focus is perpendicular to the direction of longest focus. Thus, it is perpendicular to the axis around which you need to rotate the mirror. Unfortunately, you can't tell which way to go.

Remember that you are looking at an inverted image, and, if your eyepiece is mounted to one side, you also have a rotated image. You may be unable to be sure about up vs. down and right vs. left. To determine what is the appropriate axis it is easiest to use something of known orientation near the test point. (The side of the telephone pole works fine.)

Pick the *inside* of focus line and make a slight adjustment in the objective mirror rotating it less than 1/400 of an inch around the axis defined by the *outside* of focus. This is not a rotation along the circumference of the mirror. You are rotating around a line that is a diameter of the mirror. Don't re-focus. Check the image. If it got worse (longer), you are going in the wrong direction. If it got better (shorter), you are going in the right direction.

Note: Go slow with very small adjustments. If you go beyond the correct position, it will look the same as it did before you messed with it. If you can reach your adjustment screws while looking through the eyepiece, you will be able to watch the changes. Otherwise, you might have a helper fool with the screws while you watch the image.

When you get to the point that the image of your point goes out of focus uniformly you are there!

A typical experience (with the 6-inch, f5, instrument) was that 1/20 of a turn of a 1/4 × 20 screw, thus 1/400 of an inch made a big difference.

The f9 Newtonian is very much like the f5 except for its length and the fact that the tube is made from pine rather than the oak/luan combination used for the f5. The f5 was used for the illustrations in this chapter, and illustrations of the f9 can be found in later chapters.

Chapter 7

The F17 4-Inch Refractor

This is a planetary/lunar telescope. It uses a homemade objective lens of very long focal length for its diameter, a homemade tube from oak, a 1st surface mirror for the star diagonal from eBay, and hardware for the rotating star diagonal made from PVC fixtures. Although you are unlikely to find a 4-inch f17 lens on the web there are, fairly often, 5-inch f15 s, etc.

Figure 7.1 shows the completed telescope mounted on concrete in a PVC pier. The telescope also has a tripod mounting to be used when the telescope is taken to some location other than its home. The tripod is just a larger version of the tripod for the 80 mm refractor.

This project is not entirely beginner friendly, because it uses a homemade objective lens. The reader is encouraged to consider buying a lens, but the project is included here mostly because it incorporates techniques and components that can be applied to any medium-sized long focal length refractor.

As mentioned earlier, although you are unlikely to find a 4-inch f17 lens on eBay you may find a 5-inch f15 or a 6-inch f12. The 4-inch has a focal length of 68 inches. The 5-inch has a focal length of 75 inches, and the 6-inch comes out at 72 inches. All would be comparable as far as building details are concerned. All three versions are "planetary/lunar" telescopes. By having relatively long focal lengths they can achieve rather high magnification without using eyepieces with very short focal lengths. A 3/8 inch fl eyepiece will produce 200 power that will make you feel like you are on final approach to the Moon. Add a Barlow lens and it will make you feel like you are about to land. A 3/8 inch fl eyepiece can provide reasonably long eye relief. A 3/16 will either cost a fortune or require that you jam your eye into the eyepiece until your eyelash drags on the eye lens or both.

The main tube is a 16-sided wooden tube as described in the tubes chapter. It was made from re-sawed 3/4 inch American red oak and planed – sanded to get it down to a cylinder. Except for the fact that the front bulkheads were designed

R.L. Clark, *Amateur Telescope Making in the Internet Age*, Patrick Moore's Practical Astronomy Series, DOI 10.1007/978-1-4419-6415-1_7,
© Springer Science+Business Media, LLC 2011

Fig. 7.1 The planetary refractor

to fit around the 4-inch PVC objective lens cell and the back two bulkheads were designed to provide the swivel surfaces for the rotating star diagonal the construction was standard.

The Rotating Star Diagonal

The ergonomics of a telescope and its mounting are very important. If it is awkward or uncomfortable to use it won't be used. Situations that are slightly difficult in daylight may become very difficult in the dark.

The idea of a rotating star diagonal is to allow the telescope to be used comfortably no matter where it is pointed and by a variety of different-sized individuals. The basic idea of a star diagonal goes a long way toward getting the ergonomics of a refractor reduced from pretty bad to OK. Consider what is involved in observing something at the zenith (straight up) with a "straight through" refractor. It really is not conducive to enjoyment to spend time with your neck bent at a right angle. That discomfort can be increased when the length of the telescope places the eyepiece close to the ground. Then you can sit on the ground with your neck bent at a right angle. A star diagonal does away with the neck at a right angle problem for high angle viewing. Now consider viewing an object at a relatively low angle to the horizon. Planets are commonly viewed that way. Now the user has to place himself or herself above the end of the telescope and look down into the eyepiece. The rotating star diagonal does away with that gymnastic contortion.

What we need to be able to do is allow the user to look into the eyepiece from the side. A star diagonal that rotates does away with the bent neck problem and allows observation from the side or the top.

The Back Two Bulkheads

A rotating star diagonal can be based on a PVC pipe rotating within the end of the telescope main tube. Since making a PVC tube rotate inside of another PVC tube is not practical (sticky), we need some other bearing surface. One way would be to let it rotate against wood surfaces. Two bulkheads inside the main tube with holes just large enough for a PVC tube to fit and rotate easily would be satisfactory. A slightly better solution is to take advantage of the fact that PVC fittings designed to fit over the ends of PVC pipe have tapered joints. The very ends of the female side of a PVC fitting are large enough to allow the correct pipe to enter it rather easily. Further in the fit tightens up to make the tapered joint. If you saw off the end off a female fitting 1/4 inch or a little less you get a slip ring that will go over a piece of pipe and is something like 1/32 of an inch oversize. If you buy (less than $2) a fitting intended to be a coupling you have two ends that can be sawed off.

In the example case the rotating piece was 3 inch PVC, thus, 3 1/2 inches in outside diameter. The outside diameter of the coupling is, thus, 4 inches in diameter. The PVC rings are mounted in two bulkheads about 12 inches apart. Figure 7.2

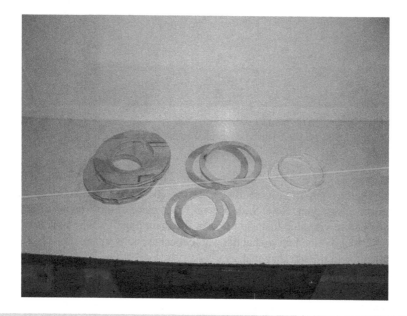

Fig. 7.2 Bulkhead parts

shows the parts used to put the rings into the bulkheads. From left to right the first
two pieces are the bulkheads with holes designed to easily accept the 3-inch PVC
but not allow the rings to pass through. The two PVC rings are shown at the right
side of the figure. The hole has diameter bigger than 3 1/2 but less than 4 inches; 3
3/4 is fine. The pieces at the bottom of the picture are a little thicker than the ring and
have a hole that will accept the ring without much motion but without deformation.
Again, 3 3/4 is good. The third piece has a hole the same size as the first piece, so
3 3/4 is fine. When these are glued together with the ring sandwiched between you
have a bearing that will allow the 3 inch pipe to move but with very little "slop."

Figure 7.3 Shown here is one set clamped for assembly, with a completely
assembled set on the right.

Fig. 7.3 Parts for the back two bulkheads to allow rotation

If the inner PVC tube happens to seize slightly on the PVC ring the ring is free to
rotate within the confines of its sandwich. Two of these, about 12 inches apart, will
give you a nice slip joint. Note: Slide a 2- or 3-foot piece of the 3-inch stuff into the
bulkheads before you glue them in place in the main tube. Use the tube to get the
bulkheads properly lined up so that the whole mechanism is square. See Fig. 7.4 for
this application to the star diagonal.

With the connection taken care of you can build the star diagonal around the
PVC pipe and develop a way of holding it in at a particular point of rotation. While
you are at it you can, if you like, arrange for the mounting of a finder scope on
the rotating part of the telescope. If you do that you will have the same ergonomic
convenience for the finder as you have for the main scope. When you rotate the
eyepiece by rotating the star diagonal the finder will go with it. If you chose to

Fig. 7.4 Rotating star diagonal

mount the finder on the main tube rather than the rotating part you should, at least, give it a star diagonal and make that rotate.

The Diagonal Mirror

Figure 7.4 shows the entire rotating part. It is made in the shape of a square tube with a 45-degree end. It is built around two square bulkheads, each with a carefully centered 3 1/2 inch hole to accommodate a piece of 3-inch PVC. On the face of the 45-degree end you can see the heads of three countersunk 1/4 × 20 machine screws. They are collimation screws for the diagonal mirror.

The diagonal mirror is mounted as illustrated in Figs. 7.5 and 7.6. Figure 7.7 shows the front of a piece of thin plywood that serves as a mounting board. The mirror rests loosely on the piece of felt and is held in place by four sheet metal clips.

Figure 7.5 shows the underside of the same piece.

The black sheet is a piece of aluminum about 1/16 of an inch thick. The two screw-heads near the middle hold it to the plywood sheet. Each screw has a couple of washers between the aluminum and the plywood. They cause the aluminum to stand off from the plywood about 1/8 of an inch while maintaining a ridged connection. The other three holes are tapped 1/4 × 20 to receive the collimation screws. Small washers cut from a junked computer mouse pad go between the 45-degree end and the mounting board.

Fig. 7.5 Underside of the star diagonal

Fig. 7.6 Adjustment side of diagonal mirror board

Fig. 7.7 Mirror side of mirror board

When the collimation screws are slightly tightened the mouse pad washers are compressed and the position of the mirror is altered. It is very important that the mirror be positioned so that its left-right orientation is square to the optical axis of the telescope. The bottom pair of collimation screws allow left-right adjustment. When the telescope is first tested it will most likely exhibit some astigmatism that will be correctable with these two screws. The third screw allows us to slightly rock the mirror back and forth for an up-down adjustment.

Screws in the plate that mount the focuser go through holes that are elongated on a fore-aft direction. This is to allow adjustment to make sure that the center of the focuser is on the optical axis. If you chose you could make the star diagonal bend the optical axis less than the normal 90-degree angle. If you wanted a 45-degree star diagonal you would set the back of the rotating piece at 22 1/2 degrees. Everything else would work out in pretty obvious ways. Bear in mind the fact that a diagonal mirror used at a lower angle needs to be of higher quality.

If you chose to use the same technique to hold the rotating part in position as was used in the example be sure that you keep the two disks lined up. A good way of accomplishing this is to attach the disk on the main tube first. Then do an entire test assembly to make sure that everything fits and can be properly rotated. You may need to make adjustments by filing one or both of the holes in the rotating parts bulkheads. It doesn't matter if you have to enlarge one hole on one side and the other on the other side. If the holes become too big to make a tight fit it is not important as long as there is a position in which everything works properly.

Apply plenty of adhesive and slide the connecting piece of PVC into the rotating part and hold it loosely in position. Slide the other end of the PVC into the main tube. Be very careful to keep it square and so that everything rotates properly without any wobble. Once the adhesive has set make sure everything still works and use construction adhesive to attach the other disk to the rotating part. Lather the two surfaces to be glued with plenty of the construction adhesive. Slide the parts together, do something to keep them properly lined up, and let the adhesive set. The reason for using construction adhesive is that it is gummy enough to accommodate small adjustments in line up. If there is 1/100 of an inch more adhesive of one side of a joint than on the other side you will still get good adhesion.

Once you have the two disks so that they stay pretty well in contact no matter how you orient the rotating part you need to provide some means of making them stay in whatever position you put them. Essentially you need to make them act like a clutch. The use of two little spring clamps as illustrated works well.

Another way of making them act as a clutch is to drill two 1/4+ inch holes in one of the rings and cut two 1/4+ inch curved slots in the other ring. Then you can use a couple of 1/4 × 20 short machine screws with wing nuts to connect the two pieces. Since the slots can't go completely around you will have only about 130 degrees of rotation, but that should be enough.

At the objective end a piece of 4-inch PVC is used for the same reason that the 3-inch PVC was used at the other end. The length of the PVC tubes helps in getting the optical components square to the optical axis of the telescope. Figure 7.8 shows the upper end with the inspection plate removed from the wooden tube.

Fig. 7.8 Objective end

Notice that the objective lens is slightly recessed into the wooden tube to provide some protection from ambient light. If we had made a small error in measuring the focal length of the objective we would be able to compensate by adjusting the fore and aft position of the tube and, thus, the objective.

Figure 7.9 shows a roughly equivalent telescope without a star diagonal. It has a simple PVC tube, and the finder is also done in PVC. It is on a very temporary mount, and the finder eyepiece is located much too far from the main telescope eyepiece.

The optics are good quality, but the telescope is very hard to use and, due to the very lightweight mounting, very shaky. Since the mounting is set into the ground it is, certainly, not portable. Accordingly there is no excuse for the flimsy construction.

Fig. 7.9 Long focal length refractor in a PVC tube

Mounting the Refractor Using a Cradle

Figure 7.1 shows the telescope that we have been describing mounted much more substantially. This mounting can also be used for the large copyscope described in an earlier chapter.

Figure 7.10 shows the cradle used to connect the telescope to the mounting. A couple of hose clamps go around the central dowel and the telescope tube to hold the telescope in the cradle. A significant advantage associated with the cradle-type

Fig. 7.10 Mounting cradle

attachment is the fact that it can be easily shifted to different positions on the tube. Just release the screws in the hose clamps and slide it.

If we wanted to attach a fairly heavy camera and projection lens to the back end of the telescope in place of the star diagonal we could simply equip the camera with a PVC tube and slide the cradle back a few inches to accommodate the increased weight.

If we wanted to convert to a straight-through design rather than the star diagonal we could slide the cradle forward a couple of inches to accommodate the reduced weight.

Chapter 8

Focusers, Eyepieces, Barlows, and Finders

This chapter includes a discussion of focusers that can be built or purchased, eyepiece types available on the Internet or which can be salvaged from other equipment, focal-length extender devices, and finder telescopes.

Focusers

The role of a focuser is to allow you to move the eyepiece in and out until sharp focus is achieved. There are four basic ways that a focuser can be designed. Focusers are almost always available on eBay and from various manufacturers if you would like to buy that part of your telescope.

There is a considerable amount of propaganda regarding what makes a focuser good or bad. You will find advertisements touting the smoothness of a focuser. If it is so rough that some positions can't be reached or maintained, then it is not smooth enough. If it operates like a washboard it is not smooth enough. Otherwise smoothness is probably overrated. It is important that it be able to reach all the positions that may be needed and that it remains where it is set and not creep. It is also important that it be able to accept any eyepieces you may want to use and that the image not move around while the observer is making adjustments. This means that the optical axis of the eyepiece must stay on the optical axis of the telescope.

How much motion is needed? The motion is generally called the "travel." The amount of travel required depends on four factors:

(1) the focal length of the objective
(2) the variation between the corrections appropriate for the vision of the users

R.L. Clark, *Amateur Telescope Making in the Internet Age*, Patrick Moore's
Practical Astronomy Series, DOI 10.1007/978-1-4419-6415-1_8,
© Springer Science+Business Media, LLC 2011

(3) characteristics of the eyepieces that will be used, and
(4) variations in the distance to the object that is being viewed.

This discussion will be clearer if we utilize some optometry terms used to specify prescriptions to correct the more common defects of the human eye.

For example, the "power" of a lens is just the reciprocal of its focal length in meters. A lens that operates like a powerful magnifying glass is regarded as strong even though its focal length is less than that of a lens offering very little magnification. Thus an objective lens with a focal length of 1/2 m (about 20 inches) is said to have a power of 2 diopters. A person can be either "near sighted" (myopia) or "far sighted" (hyperopia). He or she can also be astigmatic at the same time.

Near-sighted people see objects that are close much more easily than objects at a distance. One way of envisioning this defect is to think of the lens in their eye as being stronger than appropriate. This makes the image of a distant object reach the eye before it can get to the retina. The image is fine, but it is out ahead of the retina. So by the time the light reaches the retina the image is out of focus. The standard eyeglass correction for near sightedness consists of a lens with a slightly negative power. The prescription might be stated as –2 diopters – that is, a negative or concave lens. A near-sighted or far-sighted person can achieve perfect telescope or microscope focus without their corrective glasses by making adjustments with the focuser. If the eyepiece is positioned for perfect focus for an individual with normal sight and the near-sighted user attempts to use the telescope, he or she will have to rack the eyepiece slightly in (move the eyepiece slightly forward) to get perfect focus. Likewise, a far-sighted user needs to rack out slightly.

A 2-diopter correction requires a movement of about 3/100 of an inch with a 3/4-inch fl eyepiece. The 2 diopters is a moderate but certainly not extreme correction. A 4-diopter correction with the same eyepiece requires a movement of 6/100 of an inch. That is quite a bit of movement. As an experiment with a pair of binoculars rack them out of focus by even 1/100 of an inch and you will detect significant image degradation.

The adjustment required to adapt to a distant object depends on the actual distance and the focal length of the objective. An objective of 30 inch focal length focused on an object 100 feet distant requires an adjustment (rack out) of more than 7/10 of an inch from its "infinity" position. When focused at 50 feet the adjustment is more than 1 1/2 inches.

The important point here is that you need quite a bit of travel to accommodate varying conditions. To that you need to add a good margin for the fact that different eyepieces expect to find the primary image made by the telescope objective at different depths in the eyepiece tube. Two inches of adjustment is not excessive.

Telescoping Tubes

The simplest design for focusers is just a sliding pair of tubes. Since the most popular standard outside diameter of eyepieces is 1 1/4 inches you are looking for a tube with a 1 1/4 inch inside diameter. There is a version of a sinktrap with a properly-sized tube. One type of core tube for mainframe computer printer ribbons has an inside diameter of just a bit more than 1 1/4 inches. A piece of paper inside will provide the needed friction. The fit has to be pretty close so the eyepiece will stay put once you have positioned it. With some tubing you can distort it just a bit to get a decent amount of friction. (Squeeze with pliers, etc.) Thicker tubing can be drilled and taped from the side and a machine screw threaded in to hold the eyepiece in place. Chose a machine screw with a wing type head. This kind of a focuser will allow adjustment for the full length of the eyepiece.

Rack and Pinion Design

The rack and pinion design is the design most seen on commercial telescopes and microscopes. It has an outer tube with an inner tube that slides in and out. The inner tube (often called the drawtube) has a straight piece of metal or plastic with evenly spaced teeth attached to it (the rack). The inner tube is moved by a gear that meshes with the rack. The gear is on a shaft that is attached to the outer tube. Knobs on the ends of the shaft allow the observer to move the inner tube that contains the eyepiece. Examining any picture that advertises a telescope or microscope will provide a full understanding of this design.

Amateur construction of this design is entirely possible. Start with the gear and make the rack to match. Obviously a milling machine will help, but careful work with a hacksaw and a file will do the job. Keep the teeth as even as you can. Otherwise you will get the previously mentioned washboard effect. Look at the arrangements used in the Crayford focusers (see later) to hold the friction device, shaft, and knobs. That should provide ideas easily converted to the rack and pinion design.

Helical Design

Helical focusers are just threaded tubes. You turn one to move it inside another.

Plumbing fixtures, gas line fixtures, any pair of tubes with a male-female thread combination can be made to work. The male threaded tube should thread into the female threaded tube. The eyepiece fits into the male tube.

Plumbing and gas threads are tapered. This means that they pull up tight after just a few turns. You are unlikely to get enough travel to accommodate focusing, particularly if there are several users of the telescope, each with slightly different eyes. You can improve the situation by lapping. Apply some #220 or #100 abrasive to the male threads with a little oil and spend a few minutes turning it in and out of the female tube. It will become considerably easier to work, but you may still not have enough travel. In addition the taper of the threads may make the fit pretty sloppy at the loose end of the travel.

You can increase the travel by combining the helical idea with a telescoping pair of tubes. A problem to avoid is that of getting the eyepiece off center from the center of the focuser. You are unlikely to get anything well focused if the image moves around in the field of view as you make adjustments. In case you didn't know it, focus is critical. Most casual telescope users accept far from a perfect focus. If you switch from one observer to another, as at a star party, and the second user doesn't make any focus adjustments they are, most likely, settling for less of an image than they might have. Most folks just don't know what a truly focused image looks like.

Crayford Design

The Crayford design and the rack and pinion design have a lot in common. A fundamental difference is that the Crayford uses a friction arrangement in place of the rack and pinion. Typically the friction is provided by a rubber tube pulled over the shaft and pressed against the drawtube by springs. The knobs and shaft are pretty much the same as the rack and pinion.

Quite a number of amateurs have made good Crayford focusers mostly from PVC pipe. There are several web pages that supply excellent designs. The site http://www.atmsite.org/contrib/Holbrook/pvcfocuser/index.html is excellent. It offers a design using two pieces of rubber tubing that can be moved to adjust the friction. The instructions are very simple and complete. The site http://www.efn.org/~jcc/focuser.html is another good source and presents the actual building effort in very great detail. The site http://www.geocities.com/natsp2000/focuser.html is yet another excellent source. The designer uses both PVC and metal parts. Yet another design can be found at http://www.asahi-net.or.jp/~zs3t-tk/focuser/focuser.htm. That design makes effective use of wood and hardboard. The friction is provided by rubber O-rings.

There are commercial Crayford focusers. These are available from online catalogs and on eBay. Many allow two levels of focusing, coarse and fine.

Orion offers a basic 1 1/4 inch rack and pinion focuser for rather little money. These are plastic but work quite well. They are designed to fit the outside of a Newtonian tube. If you need to mount it on a flat surface you will have to grind or saw off two little legs. This is very easy to do.

There are two basic designs, those intended for use on refractors and those intended for reflectors. Each can be adapted to the other use. Focusers designed for refractors have flat mounting plates and, often, tubes designed to fit some particular manufacturer's refractor.

Focusers designed for use on refractors often have much more travel than those designed for reflectors. Up to a point the more travel the better. But those designed for reflectors have reduced travel because of the fundamental design characteristic that the results in parts of the tube part of the focuser being extended beyond the base of the focuser. In the case of a straight-through refractor the extra tube causes no problem. In the case of a refractor with a star diagonal the extra tube may strike the diagonal mirror, and, in the case of a reflector, this extension may go into the main tube and partly block the incoming light. If you choose to use a refractor-style focuser you may need to saw off some of the tube. Those designed for reflectors normally have some sort of bridging design to make them fit a wide variety of tube sizes.

Orion also offers a 2 inch rack and pinion for a rather reasonable cost. You can buy or make a sleeve to adapt a 2 inch focuser to 1 1/4 inch eyepieces but not the other way.

Often eBay will have several focusers listed for very low prices.

Eyepieces

Eyepieces are categorized according to their effective focal lengths. They can be made with effective focal lengths of any value from 3 inches or more down to 1/8 of an inch. The magnifying power of a telescope is determined by the quotient of the effective focal length of the objective (lens or mirror) divided by the effective focal length of the eyepiece. The shorter the eyepiece effective focal length the greater the magnification. An eyepiece with an effective focal length of 3/4 inch, used with an objective of effective focal length of 30 inches produces 40× power.

Eye Relief

Although there are design differences between eyepiece types it is approximately true that short effective focal length for an eyepiece results in smaller values for eye relief. Eye relief is the optimum distance between the back surface of the eye lens and the actual surface of the observer's eye. This optimal distance occurs when the image appears to fill the eye lens. Closer or further, and you see the object as an illuminated circle in a black tube. Short eye relief is bothersome to individuals who don't like to have anything very close to their eye and to individuals who feel that they must use their eyeglasses when using the telescope. The problems related to short eye relief restrict the utility of eyepieces shorter than 3/8 of an inch focal length. With eyepieces of very short eye relief you can feel that you have to almost shove your eye into the eyepiece to get a decent image. Eye relief is the best reason for using a Barlow lens or an erector (see later in chapter).

Exit Pupil

The exit pupil is the image of the objective produced by the eyepiece. It contains all of the visual data going into the observer's eye. The power of a telescope (a particular objective with a particular eyepiece) happens to be the number you get if you divide the diameter of the objective by the diameter of the exit pupil. Another way of expressing the same fact is to observe that, if you divide the diameter of the objective by the magnifying power provided by a particular eyepiece you get the diameter of the exit pupil. Why is that important enough to mention here? Because it imposes a limit on how low we can go in magnifying power.

The pupil of the human eye gets bigger as we adapt to the dark. The limit for a young person is about 7 or 7.5 mm (about 0.3 inches). As we age the ability to adapt to the dark is reduced because the eye's pupil just won't expand as well. A telescope with an exit pupil larger than about 0.3 inches is just going to waste light. A telescope with a 6-inch objective has a useful lowest power of about 20.

One of the value judgments for a telescope is its ability to show us very faint objects such as nebulae and galaxies. These objects don't need much magnification, so we build large diameter objective telescopes and operate them at relatively low power. We are using the telescope as a "light bucket." Telescopes used that way are what are normally called "rich field" instruments. If you chose to build a 6-inch Newtonian with a 30-inch focal length you will get the best rich field performance with an eyepiece whose effective focal length is about 1.5 inches.

What follows are some eyepiece designs that you might buy on the web or that you might build from available small lenses. There are several general types:

First there is the *single negative lens* as an eyepiece. This is the kind of eyepiece Galileo used. It is probably the earliest design, but that is about all that can be said for it. It produces a very narrow field of view and is only slightly color corrected. However, the greatest problem comes from the fact that, because it has a negative focal length, it must be positioned inside the focal point of the objective. The position needs to be as far inside as its focal length. That puts its focal point outside of the focuser range designed for normal positive lenses. The old fashioned design for opera glasses uses this type of eyepiece, probably because it is a cheap design and allows short tubes. Figure 8.1a illustrates a single negative lens eyepiece. The distance – f illustrates the focal length.

Second, there are types using only simple *plano-convex lenses*. The Huygens design uses two simple lenses. The lens nearest to the objective is normally called the field lens, which is larger than the second lens and has a longer focal length. The second lens is normally called the eye lens. For a Huygens eyepiece with an effective focal length of 1 inch you need a field lens focal length of 2 inches. With a field lens focal length of 2 inches the eye lens focal length is 0.666 inches. The two lenses should be placed with the convex surface forward and 1.333 inches apart.

Notice that the field lens is significantly further away from the eye lens than the focal length of the eye lens. The result is that the eye lens does not focus on either

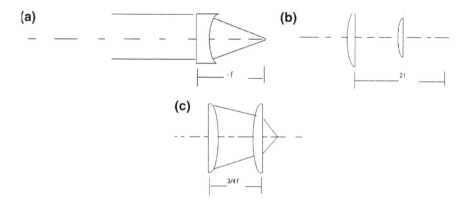

Fig. 8.1 **a** Single negative lens. **b** Huygens design, and **c** Ramsden

surface of the field lens. Otherwise dust and fingerprints become very obvious. The Huygens design is absolutely unsuitable for telescopes with fast f numbers. Used with an f15 objective lens the Huygens is a reasonably good eyepiece. Likewise it is much better at higher powers than at lower powers. The design for a higher power eyepiece can be easily scaled down from the 1-inch based dimensions. Figure 8.1b illustrates a Huygens design eyepiece.

The Ramsden design uses two plano-convex lenses with identical focal lengths. The convex surfaces facing each other are ideally separated by the common focal length. It is a good idea to mount them a little closer so, like Huygens, the eye lens is not focused on either surface. The effective focal length of a Ramsden is the same as that of the component lenses. Satisfactory Ramsdens can be made from the crown glass components of small achromats if they have become uncemented, but if you have two small achromats and want to make an eyepiece you can make a better one – a Plossel.

Achromats

The Plossel design makes a pretty good eyepiece for most situations. It is, essentially, an achromatized Ramsden. Plossel eyepieces are available from sources such as eBay in effective focal lengths from 2 1/2 inches down to 1/4 or even 1/6 of an inch.

You can expect pretty good performance from Plossels down to 1/2 inch efl. A 3/4 inch Plossel is a nice eyepiece. If you want to make a Plossel you need two small achromats with the same effective focal length and the same as the desired effective focal length of the eyepiece. Mount them just as you would for a Ramsden and set them just a little closer than the actual focal length. That way specks of dust on the field lens won't show up as boulders. A Plossel is illustrated in Fig. 8.2a.

Fig. 8.2 a Plossel, **b** Kellner, and **c** orthoscopic designs

The proportions are the same as a Ramsden. It is very important that the lenses be oriented as drawn in all of these designs.

The Kellner design can often be found on old binoculars. It uses a plano-convex field lens. With the plano side facing the objective and an achromat as an eye lens. Kellners work well on instruments with f ratios down to about 6. A Kellner can be regarded as an achromatized Huygens. It is illustrated in Fig. 8.2b. The proportions are the same as a Huygens.

Another eyepiece that may be available cheaply is an orthoscopic design. This design provides relatively long eye relief. There are many variations on the orthoscopic design, depending on types of glass available. An example is illustrated in Fig. 8.2c.

Of the illustrated types the most likely ones to be amateur built are the Ramsden, the Kellner and, most of all, the Plossel. The others are presented for identification purposes if you encounter them on some old instrument.

Very Modern Wide Angle Eyepieces

The popularity of extremely fast (low f ratio) reflectors and, to some extent, refractors has caused the development of several varieties of wide angle eyepieces that correct for the problems associated with the low f ratio objectives. These are quite expensive but may be entirely worth their cost if you are using or building an f4 instrument. It is a good idea to belong to an amateur astronomy club if for no other reason than to be able to discuss exotic eyepieces with fellow members and to try them out. Keep in mind the fact that what is needed for a f4.5 reflector may be a serious overkill for a f12 refractor. Likewise, a nice Plossel that works well on a f12 refractor may make stars at the edge of the field look like comets if you put it on a f4.5 instrument.

Compound Projection and Camera Lenses

High quality 35 mm film cameras are often replaced by digital cameras, though the optics of the old film camera may be far superior. Likewise, some of the older 35 mm slide projectors have excellent optics.

Most 35 mm camera lenses have focal lengths of about 50 mm. A 50 mm camera lens can make a very nice 2-inch eyepiece. A f2.8 Cintar from an Argus C Four is excellent. The entire C Four can often be bought for less than $10 on eBay. The C Four has its shutter behind the lens. Note that some 35 mm cameras have a shutter mounted between two components of the lens. This can make it a bit difficult to get the shutter out of the assembly and put the lens back together with exactly the correct separation and alignment.

Which side works best seems to depend on just what type of camera it came from. The Cintar seems to work well with the side that was to the outside of the camera mounted as the eye lens. A Kodak f 1.6 Anastigmat works well using the end that was closest to the film as the eye lens. There are many camera lenses that can make excellent low power eyepieces. It would be a major research project to try them all. You can get an approximation of how a lens might work as an eyepiece by putting it up to your eye, looking through it as it was being used as an eyepiece, and examining something with fine detail.

A good camera lens may perform as well as an expensive, wide-angle eyepiece on a fast (low f number) telescope.

Most camera lenses are marked with their focal lengths. This makes it easy to determine if you have a high, medium, or low power potential eyepiece. The use of a camera lens is likely to be an excellent choice for a medium or low power eyepiece.

Figure 8.3 shows a C4 lens, a Kodak Anastigmat, and a Devry f 1.6 lens each, with a piece of 1 1/4 inch tube attached to allow it to function as a nice low power eyepiece.

The biggest mechanical problem involved in building an eyepiece is that of getting the lenses square and properly spaced. If you are using an old camera lens at least the internal lenses are already properly spaced and aligned. But you still have to get it onto its tube square.

If you choose to build the simpler types of eyepiece the appropriate lenses can be had from Edmund Scientific or Surplus Sched. The 1 1/4 inch tubing can be had from Small Parts, Inc. See the suppliers appendix in this book for contact details. If you have access to a wood lathe and a reasonable level of skill using it, the appropriate spacers, etc., can be turned up from poplar. A machine lathe will allow you to do the work in brass or aluminum.

Projection lenses are another source of potential eyepieces. Many projection lenses are unsuitable as eyepieces because they have focal lengths well beyond anything that we might want for an eyepiece. Some, in fact, can be objective lenses for fast (low f number) refractors. A projection lens from a small projector may be, like a camera lens, an excellent candidate as an eyepiece. There are numerous

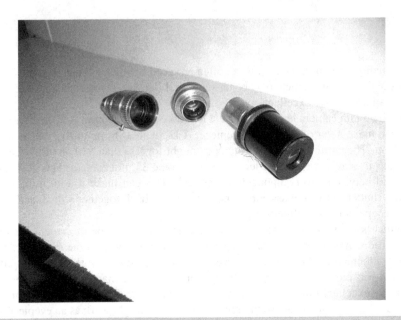

Fig. 8.3 Camera lenses as eyepieces

devices that use projection. A photo enlarger is a projector, and lenses from enlarg-
ers are excellent candidates. An eyepiece to be used for eyepiece solar projection
can well be made from a lens whose previous life was as a projection lens. As a
solar projector it is doing almost exactly what it was designed for.

Barlow Lenses

A Barlow lens is not exactly an eyepiece, but it makes sense to treat it here with
eyepiece topics. A Barlow is a negative lens used to slightly extend the image cone
produced by the objective. The effect is to cause the objective to act as if its focal
length was significantly increased. The most common examples of Barlows are designated 2× and 3×. They cause a multiplication of the effective focal length by 2
or by 3. The result is the same multiplication of the power of any eyepiece used
with that objective. The real benefit is that you can use a relatively long focal length
eyepiece and still achieve high magnification.

The step to higher magnification does not produce a corresponding reduction in
eye relief. Suppose you have an 8-inch reflector operating at f5. Its focal length will
be 40 inches. A 3/4 inch eyepiece will produce 53+ power with comfortable eye
relief. If you are examining the Moon or a planet you might want to use a 3/8 of an
inch eyepiece and get 106 power. The 3/8 of an inch eyepiece might be awkward to
use because of limited eye relief. It is much more comfortable to use a 2× Barlow

and the 3/4 inch eyepiece. You will keep the comfortable eye relief but gain in magnification.

Since nothing is free you will lose some light, some contrast, and some resolution because you have introduced two additional air/glass interfaces. Many observers, once they have discovered the Barlow, prefer it to jamming their eye into a short focal length eyepiece.

To avoid serious color problems a lens used as a Barlow must be achromatic. Remember the color errors in the "Quick and Dirty" telescope. You don't want to introduce something like that into a real scope. Except as Barlow lenses there aren't a lot of optical applications that require an achromatic, negative lens. Accordingly you may have a problem attempting to locate one in the surplus market. Complete Barlow lenses, designed to be Barlow lenses, are quite available on the Internet. The prices are often as low as $10 and seldom greater than $20, so adding a Barlow to your eyepiece collection is not a big deal. If you are doing a web search be warned that Barlow also applies to a variety of pocketknives.

Normally you use a Barlow by putting it into the focuser as you would an eyepiece. Then you put the eyepiece into the Barlow. The "front end" of a Barlow needs to be a 1 1/4-inch outside diameter tube and the "back end" needs to be a 1 1/4-inch inside diameter tube. If you have decided to use 2 inch eyepieces make the appropriate change.

Be sure you are getting one designed for a telescope rather than a microscope. The optics for a microscope Barlow might work fine, but the tube sizes won't work.

Image Inverter (Erecting Lens)

An image inverter is, for example, part of what is inside a riflescope at the narrow part of the tube. It is pretty obvious that a riflescope would be pretty useless if it produced inverted images. For an astronomical telescope we don't care, or we care very little, which side up are our images. If we are contemplating terrestrial use, even occasionally, we probably want to be able to get correct images. An inverter solves our problem and, at the same time, provides the benefits of a Barlow. It can extend the effective focal length of the objective without adding too much to the overall tube length.

A useful inverter can be made from a pair of positive achromatic lenses. Effective focal lengths of around 3/4–1 inch work well. Mount them like you would for a Plossel eyepiece. Like a Barlow, one end of the tube should fit the focuser and the other end should accept an eyepiece.

Finders

Finder scopes are almost essential to find stuff. Most of the astronomical objects of interest to the user of a telescope are invisible to the unaided eye. You can find "finder charts" for many of the interesting objects. They show the location of the desired object as it appears in a relatively low magnification finder scope.

As the aiming telescope for a larger telescope a finder scope is not really much different from any other telescope, except in its use. The construction and design criteria that apply in general apply here. The difference lies in the relationship between the finder and the main instrument.

There are two general classes of manufactured telescopes. Those that are satisfactory and those that are mostly junk. The junk cases never work very well and the satisfactory ones could perform but often don't.

It seems likely that most of the perfectly satisfactory (optically satisfactory) telescopes bought for Christmas or a birthday end up in the attic, basement, or closet after a couple of attempts because the user couldn't find anything. We know that locating something in the sky takes some real effort and practice. Even under pretty good conditions it is pretty easy to miss a galaxy because you may be looking through it or you expected something like a Hubble photograph. They aren't called faint fuzzies because they are bright and sharp. Particularly for a young person the idea that it is much harder to find something with a telescope than it is to find something without the telescope is simply counter intuitive.

A finder scope is an absolute necessity for a Newtonian telescope. If you have attempted to find anything in the sky without using a finder scope you understand the problem. The width of your little finger subtends about 1.5 degrees at arm's length. The Moon subtends about half a degree. Yes that is correct! The apparent width of the Moon is about 1/3 that of your little finger. Try it if you are doubtful. The field of view of a relatively low power (say 40 power) telescope is probably less than 3/4 of a degree. At 200 power it may be about 1/8 degree. It can take minutes to locate even the Moon, much less anything less obvious. With a straight-through refractor the finding job is a little easier. You can just sight down the side of the tube but, even that situation is less than easy. Most astronomical telescopes produce an image that is both upside down and left to right reversed. Finder-less location of bright objects such as the Moon or Venus amounts to being a matter of waving the telescope around until you detect brightness at one side of the field of view. Then attempt to follow the light. The upside down and backwards problem makes following the light counter intuitive. You have to go away to get closer! Think about moving the object rather than moving the telescope. If the object has gone out of the field to the right then pull the telescope to the left.

Requirements for a Workable Finder Scope

- Low power and reasonably wide field of view. (Happily these two requirements go together naturally.) Magnification of $5\times$ is plenty.
- Reasonable light gathering power. That tends to mean objective diameter at least 1 3/4 of an inch. (Sorry, manufacturers of 30 mm finders. They just don't cut it.)

- Ability to make fine adjustments in direction, then to lock it in.
- The finder needs to point to the same place in the sky as the main telescope. Otherwise it is useless.
- A finder needs to be easy to use. That means that the finder eyepiece must be fairly close to the main eyepiece and pointed in the same direction. Straight-through finders mounted 12 inches away from the eyepiece of a Newtonian don't cut it unless you have a rubber neck. Finders for Newtonians should have a star diagonal. (Again, apologies to some manufacturers.).
- If you have a refractor with a star diagonal then your finder needs a star diagonal. If your refractor is "straight through" then your finder should be "straight through."

Happily all the requirements are easy to achieve. There is no excuse for finders that don't work or that are awkward to use.

Satisfying the Finder Requirements

Optical quality is not really critical. A little color error or even a little distortion just doesn't matter. A lens from a good-sized slide projector can serve as an objective. A smaller projection lens can do as an eyepiece. You can even use simple lenses and tolerate quite a bit of color error.

A relatively small finder needing a star diagonal can have a diagonal mirror from a copy machine glued in place. That would be a poor idea if we were building a larger telescope. A typical small finder may have an objective with a rather short focal length. The result is that you need rather little travel in focusing. A size of 1/2 or 3/4 inches can be enough, so a couple of threaded PVC fixtures or lightweight metal pipe fixtures (gas pipe?) may do the job. You are unlikely to want to trade eyepieces between a finder and a full-sized telescope, so no harm is done if you use some oddball diameter.

Most commercial telescopes use a pair of rings with three setscrews each to hold and adjust a finder. They work well if the rings are solidly attached to the main telescope tube. These can be found on eBay and in online catalogs. Before you order, make sure that the rings will accommodate your finder and that you have figured out a good way to attach them to the main tube.

Brass pipe hangers can be used in place of the rings made for telescopes. Search your local plumbing supply store or the web. Pipe hangers provide a good way of holding the finder, but they lack the adjustment screws found on commercial finder mounting rings, so you will have to accommodate that in some other way.

Figure 6.1 shows a finder made from two projection lenses and a pretty ordinary diagonal mirror. All three of the optical components came from American Scientific On-Line. The device works fine as a finder. Part of what makes it good is the fact that it satisfies all of the criteria. Figure 6.1 shows it mounted on the f5, 6-inch Newtonian described in an earlier chapter of this book. The mounting arrangement allows adjustment both up and down and left and right. The finder is permanently attached to a flat piece of hardwood that extends about 1–2 inches ahead of the finders' objective and about 1 inch behind the finder.

The adjustment depends on a small strap hinge that is attached to the bottom of the flat hardwood piece and attached to a block on the telescope by use of only one screw or bolt. That allows an up-down movement on the basis of the hinge and a left-right movement by pivoting on the single screw or bolt. At the front end of the flat hardwood piece there is a crescent shaped slot and a machine screw with washers and two nuts. The slot allows left and right positioning, and by playing with the nuts, the up-down orientation can be adjusted. Movements are made at the front and the hinge with its single machine screw allows the finder to pivot at the back. In the illustration you will see that an extra piece has been attached to the telescope to act as a base. That piece is not strictly needed but precludes making attachment holes in the main telescope tube.

The up-down adjustment is accomplished by use of a 1/4 × 20 machine screw sticking up from the base piece and fitting through a crescent shaped hole in the flat piece attached to the bottom of the finder. The flat piece is sandwiched between a pair of nuts with washers on the stud. The small block under the hinge elevates it off the main telescope enough to clear the hinge leaf under the finder. Left-right adjustments are made by moving the finder back and forth with the crescent shaped hole moving on the stud. When it is where you want it tighten the nuts to secure the sandwich.

For attaching to the main scope, it is important that the hinge be as tight as it can be. If its blades are loose on the shaft the back end of the finder will be loose. Squeeze the hinge in a vise or hit it with a hammer (a fine optical instrument adjustment tool) to get the action tight enough to be somewhat stiff. The hinge will have to move through an arc of only a degree or less, so stiffness is not a problem. Likewise it is important that the hinge be tightly attached to the bottom of the finder. Use large "fender" washers in the sandwich and pull the nuts up tight to avoid wobble.

Figure 8.4 shows another finder, in this case, for a larger Newtonian. It also is made from two projection lenses and a first surface mirror of only medium quality. The access plate is removed to allow cleaning the diagonal. The access plate is the object under the telescope in Fig. 8.4. The mounting adjustment is accomplished by a rocker arrangement for the up-down adjustment (single machine screw) attached to a matching piece that can be swiveled on the body of the main telescope.

The rocker attached to the finder is shown in Fig. 8.5.

We have shown two easy ways to attach a finder. Take your choice or invent another. Chapter 4, describing the little 80 mm refractor, also illustrates the attachment of a finder to the main tube. In that case the actual attachment is accomplished with a hose clamp. Notice that the rocker actually looks like a rocker.

However you make the attachment it is important to have your finder mounted on the main telescope in a way that does not interfere with the balance. Do your balancing with the finder attached and with a typical eyepiece in the main telescope. The weight of those things counts. You will get less vibration and easier line up if it is balanced. It is rather disconcerting to have a telescope take

Fig. 8.4 Finder with access plate removed

Fig. 8.5 This picture shows the entire mounting and finder attached to the f9 Newtonian

a dive and end up pointing at the ground after you have located some difficult object.

In fact it is very important to get the finder and main telescope well lined up with each other. A fairly common way is to get the main telescope centered on some recognizable object (the Moon works fine), lock it in place, then adjust the finder to make it agree with the main instrument. A few minutes refining the line-up will save a lot of telescope waving. It will be much easier to get things lined up and various other adjustments accomplished if you do a trial run during daylight.

It is mostly a matter of taste rather than practicality, but you probably don't need cross hairs. Just center the object in the finder field. If you want cross hairs you have quite a variety of cross hair eyepieces available on eBay or various online catalogs. You can also buy reticules that you can add to normal eyepieces or build in to an eyepiece if you make your own. (A reticule is a piece of glass with the desired pattern etched on it.) Most of the reticules available from various online sources have special patterns. They often represent production overruns for unique applications. As long as the pattern has something that you can sight on it will serve the purpose.

If you want to convert a normal eyepiece to a cross hair type or build your own keep in mind the fact that the reticule/cross hairs must be at a focal point. For most "positive" eyepieces such as Plossels the focal point is close to and in front of the field lens. It may be a tricky job getting a reticule glued in place without getting a gluey fingerprint permanently into the field of view. If you like, and have dark hair or a dark-haired friend, you can just glue human hair in place across a short tube that fits inside the eyepiece tube. You should begin with a pretty good-sized wad of hair, because it will take some serious trial and error to get the hairs properly positioned. Keep in mind the fact that they need to be well centered. Otherwise turning the eyepiece will change your point of aim.

Don't attempt to use a reticule eyepiece for solar projection. The reticule is perfectly positioned for instant cooking. Likewise, if you are using the main telescope for solar projection put an opaque cover over the finder objective.

Peep Sight Finder

Look back at the picture of the finder on the little Newtonian. The odd-looking piece of wood with a screw-eye held to the back of the finder by a Philips head screw together with the free standing screw at the front end of the finder constitute a finder for the finder. As simple (simple minded?) as it looks, that little deal saves a lot of time. It amounts to being a peep sight such as on a target rifle. The user moves the peep around so that the screw up front fits just under whatever the finder shows in its eyepiece. Then tighten the screw. Such a little sighting arrangement may be all you need for a small refractor such as a richest field telescope. It is surprisingly precise for the same reason that such a device works for target shooting. That is, your eye will center itself on the peep automatically.

Laser Pointer as a Finder

Another approach to finder design is to use a laser pointer. Attach a laser pointer to the telescope, allowing the same degrees of adjustment as you would a telescopic finder. Adjust it to point to the same spot in the sky as the telescope. Many models are available on eBay, etc.

In use these things can eat batteries, so have spares available. The normal warnings regarding lasers apply: DON'T POINT IT AT PEOPLE. DON'T POINT IT AT AIRPLANES! The FAA (Federal Aviation Authority) has a very bad attitude toward folks who distract or blind pilots!

Chapter 9

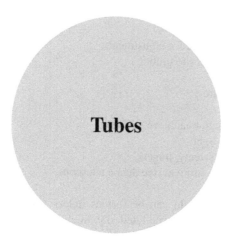

Tubes

Some sort of tube structure is required to keep the telescope optical components appropriately lined up and pointing at anything. The tube structure can be either opaque or open.

The optical parts of a telescope need to be held in pretty specific positions relative to each other, and they need to retain those relative positions when the telescope is pointed at various objects. That is a way of saying that we need a tube, or something like a tube. We are accustomed to seeing telescopes or pictures of telescopes with some sort of opaque tube keeping the front and back in their correct relative positions.

The part about correct relative positions is necessary, but the part about opaque is not really always necessary. If you look through the ads in any amateur astronomy magazine you will see lots of alternatives to opaque tubes. Aside from box shaped things (really just another version of a tube) there any many versions of open truss designs. Some are part truss and part closed tube. Others are entirely truss types, and some are, essentially, just sticks onto which the optical components are attached.

Closed tubes have both advantages and disadvantages. The advantages are

- keeps stray light out. (This is often very important.)
- strong. (Unless very poorly designed a tube will keep parts correctly aligned.)
- provides some protection. (A tube will help to keep sweaty fingers off the optics.)
- looks like a telescope.

The main disadvantage is that they tend to be heavy. This can be a real problem if you have to carry the telescope.

R.L. Clark, *Amateur Telescope Making in the Internet Age*, Patrick Moore's Practical Astronomy Series, DOI 10.1007/978-1-4419-6415-1_9, © Springer Science+Business Media, LLC 2011

Open truss designs, likewise, have advantages and disadvantages. The advantages are

- can be very lightweight.
- parts are easily accessible for adjustment.
- there are plenty of places to grab it.

The disadvantages are

- requires care in design to achieve stiffness.
- susceptible to stray light.
- susceptible to stray, sweaty, fingers.
- may look more like a carnival ride than a telescope.

Concerns about stray light can be serious unless you have a very protected observing site with a non-light polluted sky. Such sites are becoming more and more rare. Some fool with a flashlight taking a close look at your mirror can ruin your evening.

There are two basic kinds of light pollution, local and area wide. Area-wide light pollution is something you can't do a lot about. However you can, and should, join the Dark Sky Association. They work on the overall problem. Do your observing someplace else. Local is what comes from the fool with the flashlight, your neighbors' porch light, the street light out front, etc. You can achieve some protection by hanging tarps, building fences, etc. If you enclose your optics in a tube, provide good local protection for the business end of your telescope, design good baffles when you can, and make sure that the interior of your telescope is thoroughly anti-reflective you will have helped a lot. If you can convince your neighbors that they don't need to light up theirs and your backyard like a sports field, that will help. If you invite the neighbors over for an observing session while their 1,000-W floodlight is blasting away they may get the picture.

Closed Tube Approaches

Below is the variety of closed-tube types you should consider.

PVC Pipes: There is a section in Chap. 12 that discusses the use of PVC.
Aluminum Tubes: There are several Internet sources for aluminum tubes of various sizes.
Wooden Tubes: There is a large section later in this chapter that discusses two ways of building wooden tubes.
Wooden Boxes: Wooden boxes are rather easy to use. This subject is briefly covered in the discussion of PVC. Although a box-type tube may not look much like a telescope if it is used for a Newtonian it will allow close attachment of

the focuser or eyepiece. That, in turn, allows a smaller diagonal and smaller central obstruction. In some cases it is easier to attach a rectangular box-type tube to a mounting.

Sono-tubes: These are the tubes used in the construction industry as forms for concrete columns. They are available from many Internet sites but also at your local hardware store. Some commercial telescope makers use them. One of the advantages is lightness.

Fiberglass Tubes: There is an excellent Internet presentation on making your own fiberglass tubes at www.atmsite.org/Kohut/fgtube/index.htm/. There are numerous Internet sources for the glass cloth and the epoxy.

Wooden Tubes

There is something classy and friendly about wooden telescope tubes, which can't be matched by PVC. Wood is pretty light and has a low specific heat, making it quick to adapt to indoor – outdoor temperature changes. Box-shaped wooden tubes are fairly easy to find, but, after all, a box looks like a box. Let's consider octagons and beyond. These things can be made in sizes up to 12 inches and up to 16 sides. The process is fairly easy. The following procedure is the result of trial and error (make that error and trial). A small one can be done in about 4 h with a glue-setting delay (thus 2 days, about 2 h on each day).

We will describe the process in terms of octagonal shapes, but the adjustments to different shapes are obvious. The process is first described using pine and plywood. Adjustments to hardwood are noted at the end of the section.

All of the work can be done with strictly hand tools, but a table saw, scroll saw, and disk sander will make it much faster and easier. If you go the strictly hand tool route you will need to find some 1/4–3/8 of an inch pine stock. If you can re-saw by hand 3/4 stock to 1/4 or 3/8 you are a fine craftsman and don't need to be reading these instructions.

What you get looks as if it was much harder than it really is. It does require some careful and patient work, but no great degree of skill.

You probably should do the sides with solid wood rather than plywood because that will make it easier to smooth out uneven spots. You can use plywood for the sides, but, if you have to remove much wood to get the height of the sides to match, you are likely to expose an internal layer in the plywood.

Getting Started

Step 1. Lay out the shape for the ends and baffles directly on heavy paper. Then transfer the shape to the plywood that will be the ends and any needed internal supports or baffles. The plywood should be at least 3/8 of an inch thick. Even better

is 5/8 or 3/4 of an inch. An easy way to determine the dimensions is to make a circle
the size of the clear area you need in the middle. This may be the diameter of a
lens cell or a mirror cell. Draw a diameter through the center of your circle. Then,
by bisecting the 180-degree angle formed by the diameter, divide the circle into
fourths. Bisect the 90-degree angles to get eight segments, etc.; if you want more,
keep going. Using the same center as for the original circle draw another circle. See
Fig. 9.1. Allow for at least 1/2 inch of wood on all sides so these things won't be
fragile. You are going to have to either drive some fine brads or sink some small
screws into the edges, and you don't want them to shed plywood. Make your lines
good and dark so you won't lose them in the sawdust when you make the table saw
cuts and make sure you don't lose the center. You will have a compass point at the
center. Make sure it is distinctly marked. Trim off the sides to get your (in this case)
8 flats.

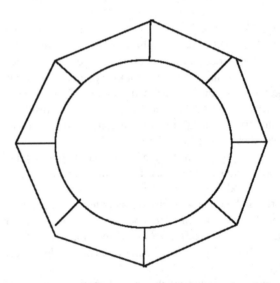

Fig. 9.1 Hexagon pattern for ends and internal supports

If you have or can borrow a disk sander do the saw cuts a little outside of the
lines and finish down to the lines with a disk sander.

You need to get all of these pieces very close to identical and get all the flats the
same size.

If you are building this thing to make a copyscope you need to cut a hole in one
piece to mount your objective lens. Exactly how you do that depends on what kind
of a lens barrel you are working with.

Obviously a barrel with threaded rings or a flange is ideal, but a completely plain barrel can be accommodated and can be held in place with a couple of automobile hose clamps, one on each side.

If you have a bare, naked lens with no barrel or cell you will have to make something to serve as a lens cell. See the section on junk collecting and examples.

Step 2. Make stock for the sides by re-sawing 3/4 inch pine boards. You can use some pretty ordinary pine shelving as long as you work around any knots. The easiest way is to carefully cut your boards to length, then rip them to width. Make them at least 1/4 inch wider than the outside width. Then re-saw them down the middle. Set the rip fence on the table saw by testing with a scrap to get the cut right down the middle. Cut them half way through from one side and finish on the other side. Obviously this limits the width of a side to about twice the blade projection you can get from your saw. Make a couple extra pieces. You will need them for test cuts.

Next, you need to saw the side pieces to width and get the angles correct. For eight sides the desired angle is 22 1/2 degrees. Set your saw table to 22 1/2 degrees and set the rip guide to remove as little of the stock as you can while still establishing the bevel on one side of each piece. Keep a few of the wedge-shaped slivers you are cutting off. You may need them to fill cracks (Fig. 9.2).

Read-adjust the rip guide to cut the bevel on the other edge. This time you need to get the width close to right to fit the flats on your end pieces. Make them a hair too big rather than a hair too small. You can always take a little off with sandpaper or a hand plane (Fig. 9.3).

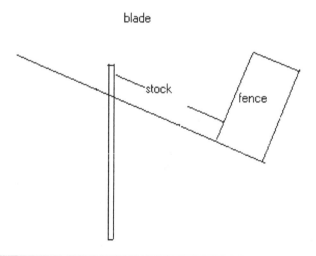

Fig. 9.2 Table saw set up for first cut

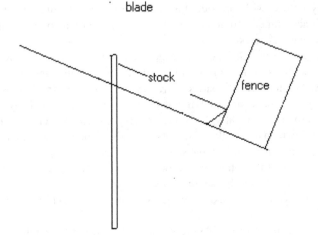

Fig. 9.3 Table saw set up for second cuts

Assembly of the Tube

The trick is to get everything square. Make sure that all your sides are exactly the same lengths.

Use a good quality, slow drying, glue and fine brads. The brads are just to hold it together until the glue dries.

Put one end piece in a vice and set a sidepiece on it so that the end of the side is flush with the outside surface of the end piece. Make sure you have plenty of glue in the joint and secure it with one brad in the middle of the sidepiece. Countersink it a little. *No more than one brad!* Reverse ends and do the same thing on the other end. It may help to have something in place to temporarily support the free end.

Now do another sidepiece, opposite the one you just did. If you think of the sides as numbered 1, 2, 3, ... 8, do number 1 first, then number 5, then number 2, then number 6, etc. Because all the sides are the same lengths, and you have kept them all flush to the outside surfaces of the ends, it can't help but come out square, or nearly so. Work reasonably quickly so the glue does not set up while you are working. Leave one or several sides not glued to the ends or baffles but glued to each other if there is more than one. Consider placing these access sides where they will eventually go using waxed paper to keep parts from being glued together where you don't want them attached. They will be attached with fine screws later. You will need to have access to the inside of the tube.

Set the almost finished tube on end, on a flat surface, and check it with a square. You will find that, because the glue is not set up and you have used only one brad at each end of each side you can push it around a little and get it to exact square.

This is important. *Tilted optical components don't work very well unless they were designed to be tilted.* Now let the glue set.

Once the glue is set you can install the baffle or baffles. Insert them through the open side and twist them into place. Glue and brad them to make them stay until the glue is set. Use just one brad in the middle of each side.

Figure 9.4 shows a completed tube for a 6-inch Newtonian. The two removable side pieces have been taken off to show the rings. The tube needs black paint on the inside and varnish on the outside.

This is probably a good time to give the entire inside a spray painting of flat black. Don't forget the unattached side.

Use a sharp plane or coarse sandpaper to even up eve the uneven places where the sides come together. You can take quite a bit of wood off and still end up with a well-rounded octagon. This smoothing step is why you used only one brad, in the middle, where you don't need to remove any wood. (You don't want to run into a brad with your nice, sharp plane.) The one brad rule also left it a little "mushy" so you could push it around to do the final squaring up.

In general, sand the living daylights out of this thing. The whole idea of doing a wooden tube was to end up with something classy. Otherwise you would have just sawed off a piece of PVC and been done with it. Once you have it nice and smooth and have removed all the re-sawing a couple of coats of spar varnish will bring up the grain and give you a finish that won't show new bumps and bruises.

Using Gussets

Two contrasting hardwoods, such as maple and walnut, can be alternated. Obviously you need a design with an even number of sides. Replace the brads with small, flat head screws.

As the number of sides increases the end pieces get closer and closer to being circles. If you want to try 32 sides you should use circles for the ends and fit the sides to the ends by gouging out the insides of the sides slightly, as shown in Fig. 9.5.

Just do the gouging at the ends. An old fashion bowl gouge would work fine. An advantage to this approach is that it would allow you to vary the width of the sides.

Following are the details of a hardwood tube for a 6-inch f5 Newtonian. The mirror came from eBay and is optical plate glass only about 3/4 of an inch thick.

The overall tube length is 31 inches The outside diameter is 7 1/2 inches. It has 16 sides. It is made from a combination of ordinary red oak and luan – the oak because it was easily available and the luan because of its rather open grain to go well with the oak. (Both have a rather coarse grain.)

A solid piece of 7/8 inch thick furniture plywood was used for the base (where the mirror cell goes) and two gussets, 16-sided on the outside with circular holes, one for the upper end and one in the middle. The sides were made in the same way as the pine tubes except that each was out of a piece of the oak about 1 3/8

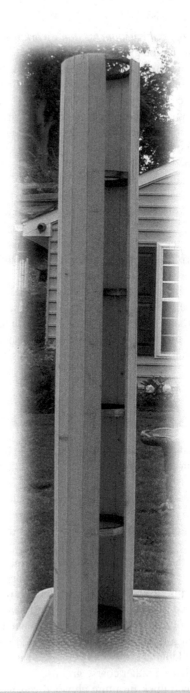

Fig. 9.4 Wooden tube for f9 Newtonian

Fig. 9.5 Gauged side

inches wide with a 3/8 inch piece of the luan glued to it to get the needed 1 3/4 inch width.

A nice "fine grain" wood pair would be cherry and either maple or birch. (A lot of birch can pass for maple anyway.) Real mahogany (not luan) and birch would be contrasting. Birch and American walnut might be nice.

Assembly requires flat head brass (because they would show) wood screws rather than finish nails. Thin pieces of oak can split if nails are driven into them. With the appropriate drilling and countersinking, the assembly takes about 5 times as long as with pine. To keep from getting it partly done and the glue setting before it gets squared up, start with four sides, opposite each other, then square it up and let the glue set. At this point the whole thing will look as if you were building some version of a lobster trap.

Even though the wood seemed well seasoned the delay while the glue set for the first set of sides may have allowed some of the unattached sides to warp. Those sides need re-sawing to get them square again. The sides were sufficiently oversized to allow for the trimming.

An effect of starting with four pieces, then working around to fill the gaps with three pieces each, is that you will have four "last" pieces to fit between existing pieces. (With the pine approach, doing piece 1, then piece 9, then piece 2 and 10, etc., you get only two "last" pieces, which have to be fit in place.) The fitting is a bit time-consuming since you have to make up for accumulated errors as you work around. Fit them in by trying each in its position and taking parts down with a block plane until you get a decent fit. It helps to attempt to plane the angles a bit too sharp rather than too shallow. You will get better actual contact at the outside of the tube where people can see the joints. A few of the wedge-shaped cut-off strips may be needed to fill narrow gaps. Don't worry about that kind of cheating. Unless you point them out the fills won't be seen.

When you start smoothing the outside you may find screws that were not countersunk deeply enough to allow safe planning and sanding. They have to be backed out. Run the countersink a couple of turns and replace the screws. Suggestion: Be a little generous with the countersinking in the first place.

Initial smoothing can be done with a block plane and may take about an hour because there will be quite a bit to remove. Some sides may stick out above their neighbors by as much as 1/16 of an inch. Keep the blade very sharp and resist the temptation to set it for a medium or deep cut. With woods like the luan and oak it is easy to lift the grain. Keep the cuts paper-thin. Note that you should not try to assemble the thing with the grain going in one direction. With so many pieces, such an effort would fail anyway. A few digs may occur in spite of the care.

The next stage can be done with coarse sandpaper in a belt sander. With it use a block and hand sanding with #80 paper to focus on bad areas.

Follow this with a vibratory sander with #100 paper. This may be the longest sanding phase. Two hours of work with more block and hand sanding may be necessary on selected areas. At this point, the digs will be gone and the surface should be pretty uniform.

A short spell with #220 sandpaper will get it ready for stain and varnish. The stain brings out the differences between the two woods. One coat of spar varnish protects the surface until the cell, finder mount, focuser mount, and attachment to the main mount are added.

Figure 6.1 (Chap. 6) shows the results. The handle is at about the balance point and makes moving the telescope very easy.

Wood Tubes Without Gussets

The internal gussets give you something to attach the sides to but require that the tube be a little larger than is really necessary to accommodate the optics. You can dispense with the internal gussets if you are willing to use thicker material for the sides and do some trickier saw work.

Saw the sidepieces just as you would for the previous version, but use the wood at the original thickness (3/4 inch). Make the angle cuts at the angle appropriate to the number of sides you want.

Now you want to cut slots in the edges of the sides. See Fig. 9.6. Make the keys (strips) to go into the slots first. Keep them all the same thickness and a little less than 1 inch wide. A piece of 1/4 inch plywood will work well and you don't have to worry about variations in thickness. Using plywood avoids the strength problem from having the grain run in the same direction as the joint. The slots need to be a little more than 1/2 inch deep and on both edges of each side piece. You can cut them fairly easily on a table saw. You will need multiple passes to get the correct width, and try each one with one of your keys to make sure you have an easy but snug fit. You don't want to have stuff all loaded down with glue and discover that something doesn't fit. That is why the slots are a little deeper than absolutely necessary.

Before you do the glue-up make sure you have several circular clamps. There are fabric clamps for jobs such as chair repair. Hose clamps can be used, but you may have to combine several to make one. Also work out something to put inside the tube as you clamp it up. Unless you are very lucky it will be something other than a circle when you fit everything together. A piece of plastic pipe wrapped with a lot of newspaper can be used to push it out to shape. It is a good idea to do a trial (no glue) assembly to make sure everything works. Glue is the only thing that will hold the thing together. There really is no effective place to put nails or screws.

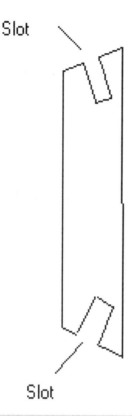

Slot

Slot

Fig. 9.6 Cutting slots in the sides

Once the glue on one of these tubes has thoroughly set you have an option not available on the gusset-based tubes with thin walls. You can set the tube up in a large lathe if one is available and turn it to perfectly round. If you make two softwood cones to stuff into the ends of your tube you will be able to get the thing on the live and dead centers of the lathe. Once you have it mounted in the lathe complete all of the turning you are going to do and do all the sanding. If you lose the center on something like this by taking it out of the lathe you will not be able to get it back.

Open Tubes

Open tubes provide the least weight of all designs but are, potentially, lacking in the rigidity needed to maintain collimation. Good open tubes can be constructed as rectangular frames or by simply drilling – cutting out parts of something that starts out looking like an opaque structure. For example, cut holes in PVC tubing.

Truss Tubes

Truss tube designs are particularly popular and appropriate for portable telescopes such as Dobsonians. Many commercial telescopes use a truss tube design. It is very important that they be carefully designed. If done with symmetrical trusses and good connections they can be very stiff.

In Appendix II of this book you will find several good references (web pages) for truss tube telescope design.

There are several suppliers for the tubing from which the trusses are normally built and the hardware to connect all the parts. Among them are MoonLite, Telescope Accessories, Markless, Scopestuff, Webster Telescope, and Texas Towers. You will find references to these and others in Appendix I.

Chapter 10

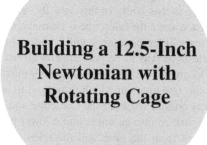

Building a 12.5-Inch Newtonian with Rotating Cage

This chapter describes the building of a reasonably large Newtonian telescope around a mirror obtained from eBay. Including shipping, the cost for the mirror, a secondary (diagonal) mirror, a mount for the diagonal, and a cell for the main mirror was less than $400.

A 12.5-inch diameter telescope is, while not a monster, a pretty large instrument. It requires careful design for the telescope and the mounting. An instrument of this size should probably not be a first project, though, with care, it could be.

A unique feature of this instrument is the rotating cage. The upper part of the main tube, containing the diagonal and the focuser, can be rotated to place the eyepiece in a comfortable location. Reasonable ergonomics requires that a minimum of twisting and balancing by the observer be required. Due to the slightly longer than normal focal length very short observers will need to stand or sit on something to slightly elevate them anyway. The 75-inch focal length forces the eyepiece to be about 68 inches above the primary mirror.

Allowing a little space below the mirror we end up with the eyepiece at least 72 inches above the ground or observatory floor when the telescope is pointing straight up. It may be worth mentioning that any object on the meridian (straight up or as close to straight up as it gets) will be at a lower elevation in a relatively short time. If you are uncomfortable on a ladder or stool, just have a little patience until your object moves off the meridian.

The intended mounting arrangement is to have the telescope permanently mounted in a shelter or an observatory. This is motivated by the fact that an instrument of this size is, of necessity, fairly heavy. Since it is heavy enough to justify permanent mounting we can proceed with little consideration of weight. As long as everything is well balanced, the added weight will just add to the steadiness. The permanent mount will be made from 2- to 4-inch pieces of iron pipe and plenty of concrete. Telescope designer Russell W. Porters description of "a mounting made of

R.L. Clark, *Amateur Telescope Making in the Internet Age*, Patrick Moore's Practical Astronomy Series, DOI 10.1007/978-1-4419-6415-1_10, © Springer Science+Business Media, LLC 2011

sand, cement and junk, with chief emphasis on massiveness." is the primary inspiration for the intended mounting. The mounting will be an inboard type, providing a great deal of rigidity by allowing for support on both sides of the main tube. This, in turn, will make a square main tube an appropriate choice. The mirror is nearly 3 inches thick and is made from Pyrex. There are various arguments for and against Pyrex discussed elsewhere in this book. In this case the added weight of the thick mirror is of no consequence since the telescope is permanently mounted. Likewise, the fact that a thick mirror will temperature stabilize more slowly than a thin one is not a problem. If the telescope was to be permanently stored in a heated area and carried to an observing location we might prefer a thinner mirror. There is a portable mounting available for the few times in which moving it to another site, such as a star party, is desired. It is possible to store the telescope and move it in four parts. The main tube can be separated into an upper and a lower part. The mirror and its cell can easily be removed from the tube, and the telescope itself can be simply lifted from the portable mounting that is a variation on the Dobsonian design.

Some of the details of the design and construction are presented here, not with the idea that a reader might use them as a plan but more as a starting point. The reasons for each decision are given to help readers make their own decisions based on their needs.

Dimensions of the Tube (Box)

The mirror came with a perfectly usable cell. The cell has three points at which it needs to be attached to the main tube. These attachment points consist of tapped holes for 1/4 × 20 machine screws located at the vertices of an equilateral triangle 14 inches on each side. The orientation of the holes is to make the machine screws run from the outside, parallel to the surface of the mirror, and pointing toward the center.

The size of the cell is greater than that of the mirror by quite a bit, making it necessary to build something pretty big to enclose it. The whole deal is fairly massive so that, assuming that the screws are run through something pretty solidly attached to the main tube, the connection between the objective mirror and the rest of the telescope will be good. The mirror cell will actually be supported on a thick plywood base with holes for access to the collimation screws. The base is recessed slightly into the main tube to protect the collimation screws while the telescope is moved. A protective cover can be fit over the mirror that, with its plywood base, can easily be removed from the tube. This is another concession to making it possible to move the thing.

Figure 10.3 shows the baseboard with one of the three clips made to attach the mirror to the base. The clips were cut from 1/8-inch aluminum bar stock. The holes for the machine screws connecting the clips to the cell are actually short slots, so the connection does not exert uneven stress on the cell. The clips are very securely screwed to the baseboard. If the baseboard is anything less substantial than 3/4 inch plywood the clips should be bolted down. Screws might pull out. You certainly don't want the mirror with its cell floating around loose in the tube.

The rotating focuser-diagonal, etc. (cage), requires the main tube be in two pieces, with the upper part rotating on the lower part while maintaining a common optical axis, a center line. If the centering is not carefully done the effect of rotating the top part of the tube will be to make the diagonal wobble. The result would be a telescope that can't be collimated. The simplest way to take care of the common center requirement is to build both parts around the same center that is in the geometrical center of both parts of the tube.

For the bottom part we need to have a square large enough to accept the mirror cell with the center of the mirror lined up with the center of the bottom part of the tube. This is not as easy a task as it may seem.

A good first step is to figure out the minimum diameter circle that can contain a 14-inch equilateral triangle and place the center of the objective mirror in the center of the circle, thus the center of the square. The minimum square will have its sides the same as the diameter of the minimum circle.

1. Draw the equilateral triangle to some convenient scale. You might even want to draw it full size.
2. Bisect each side of the triangle and connect its midpoint with the opposite vertex. The resulting lines will meet at the point that defines the center of the circle you need (You could have done two instead of three lines, but the fact that the third intersects the other two at the common point is reassuring).
3. Draw the desired circle around your center and going through each vertex. Careful measurement gives us a value of very close to 16.2 for the diameter, thus for the sides of the square that will enclose it.

Figure 10.1 illustrates a scale drafting solution. If it is followed carefully it provides a solution close enough that it will work. For different dimensions you can scale it up or down.

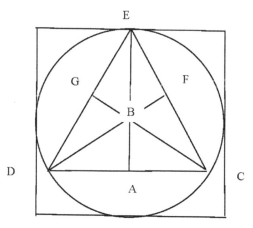

Fig. 10.1 Geometrical solution to the tube size problem

For an algebraic solution:

1. Draw the triangle and the circle that just touches the vertices of the triangle.
2. Bisect the sides of the triangle and connect each bisecting point with its opposite vertex.
3. Observe that we have made nine triangles, each of which is a 30/60/90-degree triangle. The hypotenuse of any of the smaller triangles is a radius for the circle. It is, thus, half the length of a side of the square.
4. For 30/60/90-degree triangles we have the relation 1, $\sqrt{3}$, 2 among its sides.
5. Using #3 above we have length of line segment B→C = $(4/\sqrt{3})$ × length of line segment D→C, or 16.185.

For any equilateral triangle with sides length L the sides of the minimum square that contains it and preserves the center is L × $(2/\sqrt{3})$.

See Fig. 10.1. It illustrates the 12.5 inch mirror, the 14-inch triangle, and the 16+ inch circle, all centered in a 16+ inch square. The square represents the inside dimensions of the bottom of the lower tube. The corners of the triangle indicate the positions of the attachment screws.

If you need to do this sort of design work it seems to be the easiest to work to some specific scale. Draw the square and the circle (the mirror) first, then use a cutout of the triangle to get it correctly placed. Graph paper helps, but note that it is impossible to draw an equilateral triangle with all three vertices on lattice points.

The design of the rotation arrangement requires that it not move side to side because that would destroy collimation and that it not move in and out because that would destroy focus. An easy way to avoid the in and out motion that would come from slight tipping of the cage when the telescope is aimed at a low angle is to make sure that the center of gravity of the cage is always below the rotation mechanism. This can be accomplished by a combination of some basic design ideas:

1. Have a minimum of the upper tube (the cage) extend below the level of the focuser and diagonal. In other words, let it swivel on a track just below the focuser and diagonal.
2. Devise a structure that will keep moderate pressure on the rotation mechanism to hold the two parts together.
3. Use three stabilizers to move the center of gravity down a little and to provide some safety should the top start to fall.

Figure 10.2 illustrates the overall structure of the tube.

The 16 1/2-inch by 16 1/2-inch bottom plate is cut from 3/4-inch plywood, with fairly large holes cut for air circulation and access to the collimation screws. The sides are cut from 3/4-inch plywood for but joints. So the sides are actually 18 inches wide and the front and back pieces are exactly 16 1/2 inches wide. The two back corners are gusseted with 45-degree molding. The molding at the back corners is relieved to accommodate the bottom and middle plates.

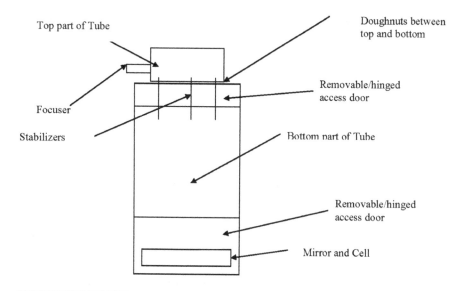

Fig. 10.2 The overall structure of the tube

The bottom plate rests on, and is screwed to, 2 by 2 rails attached to the back and both sides of the box. The screws can be removed to pull the bottom plate, cell, and mirror out. The bottom plate is actually raised 3 inches from the actual bottoms of the plywood sides. This allows space for some reinforcement pieces and a lightweight dust cover. The whole arrangement allows enough space so the tube can be safely set on its bottom end without there being any unwanted pressure on the collimation screws. The front side of the box has a removable section allowing access to the mirror and its attachment screws.

The height of the sidepieces of the bottom tube can be calculated as follows:

At the bottom we need to account for:

Bottom of cell to mirror surface	4 3/4 inches
Collimation screws extend below the bottom of the cell	1 1/4
Space to protect the collimation screws	2
Thickness of plywood	3/4
Total	8 3/4 inches

The surface of the mirror is 8 3/4 inches up from the bottom of the plywood sides. The primary image will be formed 75 inches above the mirror (fl = 75 inches).

The top part of the tube will be a box, like the bottom part, but it will be only 14 inches on each side.

To direct the primary image out the side of the upper tube we use the following:

Center of cell to side of tube	7 inches
Thickness of side	1/4
Focuser height	4
Height of upper tube below focuser	3 inches
Total	14 1/4 inches

75 inches + 8 3/4 inches – 14 1/4 inches = 69 1/2 inches

The bottom part of the tube is 69 1/2 inches high. It is strongly braced at the top and is partly closed by a second 16 by 16 piece of 3/4 inch plywood. At a position somewhat below the middle there is a third piece of 3/4 inch plywood with a 13-inch hole. This is to serve as a light baffle and to provide additional stiffness at the point where the supports are attached. The closing plywood has a 13-inch diameter hole around which the rotation track is mounted. The top section of the tube is the "cage" for the diagonal and the focuser and is a box 14 inches on each side. It must rotate on the 3/4-inch plywood piece, which terminates the bottom part of the tube. This whole assembly must be built to be very square and to stay that way!

The sides of the lower box are made from 3/4-inch plywood and are gusseted at the corners by triangular pine moldings.

The main mirror, the diagonal, its mounting, and the stalk spider are illustrated in Fig. 10.3.

Fig. 10.3 Mirror, cell, baseboard, diagonal, stalk, and stabilizers

This figure shows the mirror in its cell. The cell is attached to the baseboard with three aluminum L-shaped brackets. The stalk, attached to the diagonal, and the three stabilizer pieces are laying next to the baseboard. The aluminum brackets are easily cut from fairly thick aluminum bar stock and drilled prior to bending. To make line up easy use slots rather than holes for the machine screws. That way you will have space for a little sliding around.

An additional piece of 3/4-inch plywood goes in the main tube at the absolute bottom. Its purpose is to allow a hand-truck to be used to move the completed telescope. Without this additional protection the lip of a hand-truck, when slid under the tube could, when the tube is raised, make contact with the collimation handles. The additional piece is a copy of the real baseboard with holes for access to the collimation handles and the locking screws.

An appropriate next step at this point is to give the interior a coat of flat black paint or glue in black flocking material. You need to get the non-reflective paint or flocking on every surface that can be seen from any point on the mirror or the eyepiece. After the paint is dry the front parts can be installed.

Your design will, most likely, involve different numbers and different criteria, but you will need to do some sort of calculation like the above. Be careful to take all the critical distances into account. An obvious sign of a Newtonian telescope for which the arithmetic was done quickly is a "moved" hole for the focuser. The moved diagonal may not show, but it is hard to move a hole without leaving a trace.

The top part of the tube needs to extend above the level of the diagonal and focuser by enough to serve as an effective block for incidental light. Since the top will be removed on the rare occasions that the telescope will be moved length is not a serious issue. To be on the generous side the top was designed to have a total height of 15 inches above the rotation track. That allows for 3 inches below the focuser/diagonal and 12 inches above to block incident light.

The bottom of the upper part of the tube must hold the mechanical parts that mate with the rotation track. They are mounted on yet another piece of 3/4-inch plywood that fits in the bottom of the upper part of the tube in the form of a doughnut.

Diagonal

The diagonal mirror supplied by eBay with the primary had a minor axis of just a little less than 2.5 inches. Examination of the table of diagonal minor axis values based on primary size, focal length, and distance from the diagonal to the primary image gives a value of 1.938 inches for a 12 inch primary with a 72-inch focal length and the primary image projected off the side 9 inches. In this application the primary is a little bigger, but the focal length is a little longer. We can assume that those two approximations cancel each other out. The distance from the center of the 14-inch upper tube to the primary image will be more like 11.25 inches. The 11.25 inches comes from:

Center to inside edge of tube	7 inches
Tube wall thickness	0.25 inches
Distance into the focuser to the focal point of the eyepiece	4 inches

The extra distance of about 1 inch increases the size of the appropriate diagonal slightly. In conclusion, the supplied diagonal is just a little bigger than necessary but not enough to be any problem.

The diagonal mirror came with a 45-degree mounting having a 3/8-inch threaded stud on the back. The 45-degree mounting has four adjustment screws to make fine adjustments to the position angle of the diagonal mirror. Experience indicates that screws close to the diagonal are difficult to access from the outside of the telescope when the diagonal has been mounted in the center of the tube. Perhaps the most serious problem associated with screws inside the tube is the danger of dropping a screwdriver. You can pretty well figure that it will hit blade down somewhere on the main mirror. You probably can't completely eliminate internal adjustments, but reduce them, keep them easily accessed, and, when you have to use them, keep the tube as close to horizontal as you can.

The diagonal, it's mounting, the stabilizers, and the stalk spider are shown in Fig. 10.3 with the baseboard and the mirror.

Spider

The purpose of a spider is to position the diagonal in the center of the tube. Between the diagonal mounting and the spider it is absolutely necessary that you have the ability to adjust the position of the diagonal in three dimensions. In addition the angle of the diagonal must be completely adjustable. Not only must these adjustments be available but, once made, they should not be subject to slippage.

An easy way is to build a one-legged spider or a stalk. A more typical telescope spider has three legs. Some have four legs. We are using one with only one leg, but considering that a real spider has eight legs we are not doing too much damage to the term "spider."

This spider is specific to the case of a square tube. It is attached to the side of the tube with two bolts. Adjustment of the distance from the tube wall to the diagonal is accomplished by bending the stalk. Notice in Fig. 10.3 that the bend in the stalk is less than 90 degrees. If the bend is adjusted closer to 90 degrees the distance from the wall to the diagonal is increased. If the bend is adjusted to further from 90 degrees the distance is reduced. To keep the diagonal centered in the other direction the base of the stalk, where it is bolted to the side of the tube, needs to be moved. The bolts are run through slots rather than simple holes. The slots could have been made in the base of the stalk or in the side of the tube. The choice to avoid removing extra metal from the stalk is based on a concern for weakening it. The stalk will be bent to adjust the angle. There is no point in weakening it near the bend point. Aluminum won't take a lot of bending. Some of the collimation adjustments will involve reaching down the throat of the beast and bending its glottis (the stalk).

Upper Tube Rotation

The upper tube has to be able to be rotated on the lower tube to position the focuser-eyepiece in a convenient position no matter where the telescope is pointed. The most important criteria for the rotation arrangement is that the collimation of the telescope is not disturbed by the rotation process. This requires that:

1. The axis of rotation must be exactly the optical axis of the telescope. Essentially this means that the diagonal must not move laterally when its cage is rotated. The diagonal must be very well centered in the tube, and the tube must rotate around its center. If we did not have the rotation requirement slight errors in centering could be compensated for by tilting the primary mirror. If the diagonal was off center by 0.1 inch it would be off the optical axis by about arctan $(0.1/75) =$ arctan $(0.00133) =$ about 5 min of angle. By tilting the objective mirror about 2.5 min of angle we would have a perfectly workable compensation. Given our rotation requirement the compensation would have the effect of doubling the error when the cage is rotated by 180 degrees.
2. There must be no lateral wobble when the cage is rotated. When the cage is rotated the diagonal – actually the line of sight from the eyepiece reflected by the diagonal – must not wave around. A laser beam from the center of the eyepiece to the apparent center of the diagonal must hit the objective mirror in the center or VERY close to there. This is the hardest criteria to satisfy, but it is the most important, since it is equivalent to requiring that collimation be preserved independent of rotation.

A way of satisfying these requirements is described below. There are, certainly, other and perhaps better ways, but this is pretty easy and works pretty well. It is a scaled-up version of the rotating star diagonal used on the 4-inch f17 refractor.

Doughnuts to Preserve Alignment

Make two plywood doughnuts about 28 inches in diameter with 13-inch doughnut holes. These should be made from very high quality plywood (minimum 3/8-inch thick) or hard Masonite. Use a material unlikely to warp. The 13-inch holes must be well centered in the 28-inch circles. If you make one of the doughnuts 28 1/2 inches in diameter it may look better, since they are going to be stacked. By the time they are sanded, if both start out the same size, they may not appear to line up very well. This is just a cosmetic consideration, so it is only as important as you want it to be.

Now you need something to work as a bearing surface. It has to be fairly slick, so rotation is easy, but it need not be extremely tough. It is not as if you were making wheel bearings. These things will be moved only a few times each evening.

White Masonite Wall Paneling. Intended to serve as paneling for areas subject to moisture, it is about 1/8 of an inch thick. Shiny white on one side and Masonite brown on the other, the white surfaces slide well, one over the other, and can be lubricated with a little silicon grease or some such lubricant. The stuff is only moderately stiff, so it needs to be backed up by solid plywood or some such material.

Georgia Pacific Fiberboard Wall Paneling. Also intended to serve as paneling for an area subject to moisture, the surface is very lightly embossed and has a rather unattractive pattern in reddish brown. It probably comes in a variety of colors more attractive than this. That may be why a 4 ft by 4 ft sheet was available for $5. In the telescope application the unattractiveness is of no importance. The light embossing is not enough to cause any unevenness when one surface slides over another.

You might chose to use either of the above, cheaper materials, or something else with equivalent characteristics. The important characteristics are that it be reasonably slick, that it can be easily cut to shape, and that it have a reasonable degree of toughness. You are going to use it to make two more doughnuts. These are to force and maintain precise alignment between the top and the bottom parts of the tube.

Doughnut 1	Outside shape diameter	Round to fit the plywood doughnut on the top of the bottom section of the tube. Thus about 28 inches in diameter
	Inside shape diameter	Round, 19-inch diameter hole
Doughnut 2	Outside shape diameter	Round, 19-inch diameter. If you use an extremely fine blade to cut the 19-inch hole in Doughnut 1 and keep the cut very circular you may be able to use the cutout from Doughnut 1 for Doughnut 2
	Inside shape diameter	Round, 13-inch diameter hole

All of the cutting, including the hole cutting, can be accomplished with a reciprocal saw. Use a fine blade and work carefully. Be particularly careful to get Doughnut 2 to fit into Doughnut 1 with a minimum of slop. Figure 10.4 illustrates both doughnuts in the order that they will be assembled.

As we said above, shapes like these can be made with a reciprocating saw, but if you have or can borrow a router with a straight bit it is easier to get nice neat cuts. There is a web page providing a very good explanation of the circle cutting process. It is www.woodworkingtips.com/etips/etip102000sn.html. If you chose to cut with a reciprocating saw you need a way of drawing good circles on the material with a pencil. You can make a pretty good compass by selecting a piece of scrap plywood longer than the desired radius. Near one end drill a hole that provides a force fit for a pencil. Sharpen the pencil to have a stubby point so that it will be held by the hole with only about 1/2–3/4 inches of pencil sticking out. Measure from the middle of

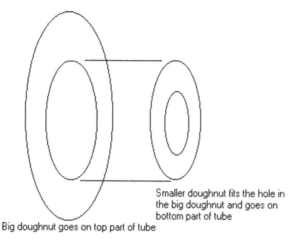

Smaller doughnut fits the hole in
the big doughnut and goes on
bottom part of tube

Big doughnut goes on top part of tube

Fig. 10.4 The doughnuts

your pencil hole a distance equal to the radius of the desired circle. Drive a nail through the plywood at that point. This is your compass. Drive the nail through the material to be cut and run the pencil around. The nail hole will define the center to use for any circles concentric to the first.

The larger doughnut will be glued and clamped to the plywood doughnut on the bottom of the upper tube. The smaller doughnut is to be glued to the top of the lower part of the tube. Be very careful about centering these parts. Their positions could just as well be reversed.

Suppose that Doughnut 2 is a little loose in Doughnut 1 or that a little wear occurs allowing slight lateral motion. What to do?

Assuming that one or both of the inside of the large doughnut or the outside of the small doughnut are pretty close to circular you can add a little to either edge. Add the same amounts to the edge in three equally spaced locations around the circle. For example, 5 or 10 fine brads with their heads removed can be driven in a line. Get them down to just below the surface of the doughnut. The other circle will now have three new points upon which to ride. Try to get smooth surfaces so the circle resting on them does not abrade rapidly.

Stabilizers

Stabilizers are pieces of soft steel bar cut to about 10 inches length, with two 5/16-inch holes drilled at one end and a right angle bend applied to allow them to be screwed to the "floor" of the upper tube and extend about 8 inches down into the lower part of the tube. They are shown as simple vertical lines in Fig. 10.2.

They should be spaced about evenly around the 13-inch hole and can be bent slightly outward to make it necessary to compress them slightly to remove or replace the upper tube. They are just long enough to catch on the hole in the lower tube if the upper tube starts to tip off. They also contribute a little weight to lower the center of gravity of the upper tube.

Mounting

The Dobsonian-type mount is illustrated in Fig. 10.6. The balance of the whole telescope is very critical. The pivots on the side of the main tube must be located very close to the center of gravity, or an excessive amount of friction will be required to keep the telescope pointing at an object. The next chapter provides an extensive discussion of mount types. It is pretty obvious that a Dobsonian mount is really an altazimuth version of the standard fork type.

An easy way of balancing fairly large telescopes like this one is as follows:

1. Get everything in and on the telescope that is going to be there. This includes the main mirror, the diagonal, the focuser, a typical eyepiece, etc.
2. Set the whole thing on its side on a flat bench (Depending on weight you may need help).
3. Set it on a broomstick or equivalent dowel and keep rolling it around until you find a point of equilibrium.
4. Mark that point. This is where you need to pivot the whole thing.

Fig. 10.5 Simple dobsonian mount

Fig. 10.6 The 12.5-telescope on the dobsonian mount

An easy and functional way to allow use of a Dobsonian-type mount but reserve the ability to use a fork or yoke for a permanent mounting is to attach iron pipe flanges to the sides of the tube at the point of balance. Screw pipe nipples into the flanges to serve as the bearings in a Dobsonian-type mount, but attach better bearings, gears, etc., to the pipe nipples for a more elaborate permanent arrangement.

For this particular telescope the Dobsonian mount is very rudimentary, since it is only seldom used. It is simply a three-sided box supporting the pipe nipples about 39 inches above the bottom of the box. A pipe hanger is used to act as a clutch. The bottom of the box sits on a bearing of shiny Masonite lubricated with grease. The pipes sticking out from the sides at the balance point provide handles for lifting the thing onto the Dobsonian mount if there is another person available and for attaching a lifting strap for use by one person with a small mechanical lift.

The permanent mount is not illustrated. It is a yoke design that is really just a fork with the upper end elongated to accept the entire length of the telescope and closed so that the equatorial axis can be supported at both ends. Figure 10.5 shows the mount and Fig. 10.6 shows the telescope mounted on it.

Chapter 11

Ergonomic Mounts

Any telescope needs a mount. It just doesn't work to wave even a small telescope around and hope to see anything. There are numerous fairly easily built styles and many complete mounts available from Internet sources. The most important characteristic of mount, after steadiness, is comfort. Ergonomics are important.

Mounting Types

There are two basic types of telescope mountings, equatorial designs and altazimuth designs.

An altazimuth type mount allows motion in all directions, somewhat like a ball joint. As a matter of fact, a ball joint like those used with camera tripods can be used as a part of an altazimuth mount.

A Dobsonian telescope is just a Newtonian telescope on a very ingenious and ergonomic altazimuth mount. It can be thought of as an altazimuth version of a fork type equatorial mount.

An equatorial mount has two axes at right angles to each other. One axis is called the polar axis. The other is the declination axis. The polar axis must be parallel to the rotational axis of Earth, and the declination axis must be at right angles to the polar axis.

There are several general designs for equatorial mounts presented in earlier chapters of this book and Part III of *Amateur Telescope Making Volume I*. One of those designs is called an English or fork design. A variation of that design can be constructed from wood and a few pipe fittings. An earlier chapter includes two examples of such equatorial mounts. One holds the long 6-inch Newtonian and the other holds

R.L. Clark, *Amateur Telescope Making in the Internet Age*, Patrick Moore's
Practical Astronomy Series, DOI 10.1007/978-1-4419-6415-1_11,
© Springer Science+Business Media, LLC 2011

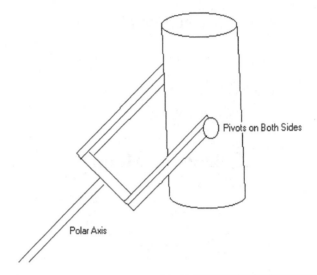

Fig. 11.1 General fork mounting

the shorter version. The design is obvious from the picture. Steadiness is important, so don't get cheap on the pipe sizes (Fig. 11.1).

Using an altazimuth mounting is more intuitive than using an equatorial mounting, because the telescope can be moved from any pointing position to another directly. The advantage of an equatorial mounting comes from the fact that an object can be followed by rotation of only the polar axis. For example, the normal motion of the sky moves an image of the Moon across its own width in 2 min. Finding things in the sky, even such obvious stuff as the Moon, can be a seriously frustrating job. If you are using an altazimuth mount and you allow an object to move completely out of the field of view you will have to repeat the finding process. If, instead, you are using an equatorial mount you only have to nudge the telescope to the west to catch up with a lost object.

Supports

Another design variation among mounts is based on just what supports the thing. A tripod is a common solution. Tripods are often used with mounts intended to be portable. Solid piers are a common solution for permanent installations. This is not a hard and fast differentiation, because there are portable piers and there are permanent tripod installations. Even a portable pier is likely to be rather heavy, and tripods, portable or otherwise, have legs sticking out to be kicked and/or tripped on.

There are designs that are neither tripod nor pier supported. The classic, and easily built, pipe mount sits on a wooden triangle, and the Dobsonian mount sits on a sort of pier only a couple of inches high.

Balance and Moment

If the telescope is to be easily pointed at something in the sky and to stay where it is pointed we want the combination of the mount and the telescope to be in balance. More exactly, we want it to be in balance no matter where we point it. An arrangement like a pendulum can have a configuration in which it is in balance (The weight hanging straight down). If it is moved off its balance point it will return to the hanging position. To get our discussion of balance stated precisely enough to allow us to test a proposed design without having to actually build the thing and test we need to introduce (gently) the concept of torque.

Torque involves a force acting on a lever arm and a pivot point. It is measured by the formula:

$$T = r \times L$$

It comes out in units like foot-pounds or gram-meters. L is the length of the lever arm, and r is the force.

In the formula we assume that the force – the pounds or grams – is applied at right angles to the lever arm. If the force is something like gravity, which it is in our case, then when the force succeeds in moving the lever arm the angle will be changed. This is why careful engineers define the force as a vector. Vectors combine the idea of magnitude (force) with direction. The fact that torque is a linear function of the length of the lever arm is illustrated by the fact that a stuck screw is more easily moved by a long wrench (spanner) than by a short one. More easily broken as well!

The formula explains, and is illustrated by, the situation of a 150-pound man and a 75-pound child on a teeter-totter. The child must be exactly twice as far from the balance point as is the man.

Our "balance" concept is more precisely stated as: "We want all torques to add to zero around our pivot point." The point in an object at which all the torques caused by gravity add to zero is called the "center of gravity." The pivot point is the axis of movement of the telescope and the moving parts of the mount.

In Fig. 11.2a we have a refractor without a finder. It is mounted with a pivot point along the actual center of the telescope and positioned so that, when the telescope is in a horizontal position, everything is nicely balanced fore and aft.

In Fig. 11.2b the telescope has been equipped with a finder. Suppose it is elevated to point at something in the sky. In that orientation it is out of balance and is heavy on the eyepiece end. What has happened is that the vector representing the weight of the finder has shifted back toward the eyepiece so that the moment of the combination

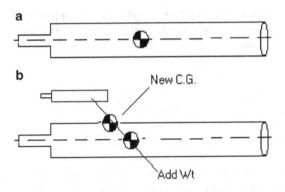

Fig. 11.2 a, b Balance problem

no longer goes through the pivot point. Another way of expressing this is to say that, because the center of gravity was above the pivot point, tilting the telescope upward shifted the center of gravity. Initially the center of gravity was directly above the pivot, and we were at a position of very local stability. As soon as we moved the telescope it became like a person standing up in a canoe. Once they get the center of gravity of their body, together with that of the canoe, out over the water they end up in it!

The balance condition that we really want is for the center of gravity to actually coincide with the pivot point.

An obvious solution to the problem illustrated by Fig. 11.2a, b is to raise the pivot point by enough to get it to coincide with the true center of gravity of the combination of the telescope and the finder. A new pivot point would solve the problem. Another solution, not so elegant, is to add some weight in a position exactly opposite the finder. Added weight at that position would do it. In this case "opposite" means, "with respect to the pivot point." The line from the center of gravity of the finder, through the pivot point, defines where "opposite" can actually occur. The closer to the pivot point the more weight will be required to generate the same moment as is generated by the finder.

It is generally better to get the pivot in the right place than it is to add weight to move the center of gravity to a pivot point. If you have to adapt an existing mount to your requirements you may not have a choice. If you have to work with a fixed pivot point but need to accommodate something like an overweight finder you can weigh the new item and calculate the new moment that it generates (distance from the pivot point to the center of gravity of the new item times the added weight). To counterbalance the new item just add an opposite moment. In general the center of gravity of a complicated object can be determined as the center of gravity of the several centers of gravity in different parts of the object.

The problem illustrated by Fig. 11.2a, b actually occurs with many commercial mounts. It is not at all unusual to find an arrangement with the telescope, a standard eyepiece, and a finder supplied with the telescope very well balanced on a mount supplied with the telescope. If the user attaches a larger finder, uses a heavier eyepiece, or attaches a camera the balance is disturbed. Commonly the telescope is supplied with some way of attaching a counterweight or a movable weight, but the attachment may be in an inappropriate location.

Most mounts, homemade or otherwise, have a certain amount of friction either intrinsic to the bearings or supplied by some sort of clutch or brake. Minor imbalances can be accommodated in that way, but if there is more than 1 or 2 foot-pounds of torque these arrangements may slip or induce vibration. A device normally used as a pipe hanger to attach water pipes to a ceiling has a clamp operated by a screw that is intended to grab the pipe. It can be used to grab a shaft and can serve as a clutch. You can replace the tightening screw with a threaded rod and a handle to get a pretty good grip on a shaft. These gadgets are available from online hardware suppliers or plumbing supply organizations.

The weights from an exercise weightlifting set make excellent counterweights. Cast iron (the most common material for such weights) has a specific gravity of more than 7, while concrete has a value of about 2.3. That means that it takes about 3 times as much concrete as it does iron to produce a counterweight.

An old-fashioned, fairly easy to construct mount, particularly for Newtonian telescopes, is commonly called a pipe mount. It is constructed from plywood, some 2 by 4 s, a couple of U bolts, two weights, stray fasteners, and various iron plumbing parts. The diameters should be at least 1 1/2 inches. Incidentally, plumbing stuff is rated according to the inside diameter.

The following is a parts list for what you will need:

One piece of plain pipe about 2 or 3 feet long. Threaded at least on one end.
Two flanges
Two short pieces of pipe threaded on both ends (Called nipples).
One "T" joint
Two weights
A small amount of pretty thick plywood

Figure 11.3 illustrates a pipe mount permanently mounted on a couple of 4-inch by 6-inch timbers. The row of holes in the sides of the head allow adjustment of the position of the weights. A portable version would be on a V-shaped structure, made from 2 by 4's. From the point of the V a 2 by 4 would angle upward and support the pipe. The angle should match your latitude. The whole thing should face north.

The web page www.charm.net/~jriley/sky/mount1.html provides excellent plans for a pipe mount designed for portability. This design also allows the use of setting circles. The web page www.astro.ufl.edu/~oliver/ast3722/lectures/BasicScopes/BasicScopes.html provides very good mounting ideas.

Fig. 11.3 A permanent pipe mounting with the 6-inch long (f9) Newtonian

Wedges

A wedge is a commonly used component of a mount. It is an adjustable steel or aluminum construct that can be attached to a tripod or a pier and provides a solid transition from a level surface to a correct polar orientation. If you live at 39 degrees north latitude you set the wedge at 39 degrees. If it is your intention to build a mount from parts such as a tripod from one source, a geared head from another, and a wedge from somewhere else you need to be sure that all the parts will fit together. The reason for the existence of a wide variety of adaptors is that many of these things are not completely compatible. If you can, visit some retail outlets and attend some star parties so you can actually look at the variations offered by various manufacturers. Ask questions of online sellers such as whether a certain head fit a certain wedge.

The mounts available on the web include new and used versions of every type ever made. Unfortunately there is not much in the way of comparability. The web page for Astromart Classifieds has numerous listings for mounts and parts of mounts.

An Ergonomic Altazimuth Mount

A comfortable altazimuth mount for a refractor has no legs in the way and is capable of pointing in any direction, including straight up. It is also easily adjusted so observers of any height can comfortably get to the eyepiece (Fig. 11.4).

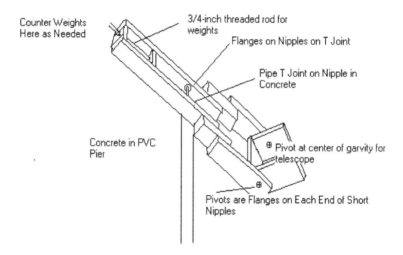

Fig. 11.4 Ergonomic altazimuth mount

An obvious ideal mount design would support the telescope by some sort of zero gravity magic. The user would just push the telescope wherever he or she wanted it and expect it to hang there. With this mount we get as close to the ideal as possible without the use of zero gravity magic.

The basic idea of this mount is to use a "U"-shaped cradle-like support for the telescope up the sides, which makes the telescope and the "U" sit at perfect balance. The "U" is swung between two connected beams. The beams are supported by a "T" pipe joint, allowing the whole mechanism to rock up and down and to rotate around a vertical axis. The whole deal may look like one of those big pumps found in oil fields. The only counterweight required is at the opposite end of the beams.

A previous version required two counterweights. Changing telescopes was a problem because it required that both counterweights be altered (Figs. 11.5, 11.6 and 11.7).

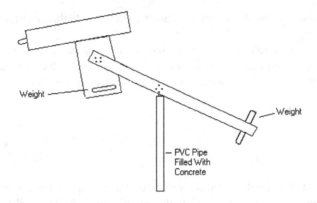

Weight

Weight

PVC Pipe
Filled With
Concrete

Fig. 11.5 This needs adjustment of two weights to change telescope

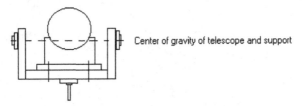

Center of gravity of telescope and support

Attach an extra pipe flange and stub of pipe at
the bottom as a handle.

Fig. 11.6 Cradle for telescope

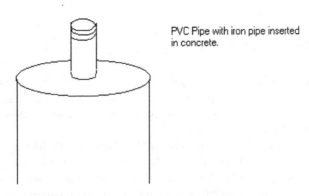

PVC Pipe with iron pipe inserted
in concrete.

Fig. 11.7 Top of PVC and concrete pier

A note about pipe joint bearings. Pipe joints are designed to be tightened very solidly. Otherwise they would not work for the designed purpose. They do that by using slightly tapered threads. To get pipe joints to work well as bearings you need to "lap" them. Use a very little #120 abrasive or valve-grinding compound with some light oil (WD40) in the joint and just twist them in and out for a few minutes. You will be able to pull them up by about two more turns and you will get a much smoother bearing.

There is always the slight danger, with the connection under the "T" joint, that you will completely unscrew it. Always begin an observing session by tightening it up until it almost binds. Then you have 8–10 complete revolutions before it gets loose. You certainly don't want the whole deal to fall off the pedestal. It is pretty heavy.

Some additional thinking about the mount. When showing an object to another person such as a child the difference in peoples' heights is often a problem. You may locate some neat object, get it centered in the view, and discover that the eyepiece is too high or too low for the other person. By the time you get the ladder or stool positioned the object has moved out of the field of view. A parallelogram arrangement such as is used for some binocular mounts might be designed so that the telescope would stay pointed in the same direction when pushed straight up or down. The trick to designing such a thing is to keep the opposite sides of the parallelogram equal in length and to avoid slop in the joints.

A very popular version of the fork-type mount is the Dobsonian. The mount shown for the 12.5-inch Newtonian in the previous chapter is a simple Dobsonian. There are several excellent web pages on Dobsonians listed in Appendix II of this book.

Chapter 12

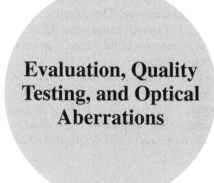

Evaluation, Quality Testing, and Optical Aberrations

This chapter provides corrective actions if your telescope fails to perform as well as you think it should and includes a discussion of optical faults and testing techniques that are relatively easy to apply. Descriptions of the common aberrations are supplied with an explanation of their effect. A section on quick and dirty quality tests is included.

How Good Does It Need to Be?

The demands on an optical system to be used on astronomical objects are much higher than for other uses. A telescope that works fairly well for tourist-type situations may be a disaster when you point it at a star. Nevertheless, the first quality object test for most instruments will be a distant tree.

Daylight testing can do quite a bit, and it is a good way to get accustomed to the process of aiming and focusing. You can construct an artificial daylight star by hanging a Christmas ball decoration a few hundred feet away, where you can see its reflection of the Sun. However, above all, DO NOT TRY TO TEST THROUGH A WINDOW. As optical instruments windows are terrible. In addition to problems from irregular surfaces they generally can introduce severe astigmatism. The more modern, energy saving windowpanes have a partial vacuum between the panes. That makes each side distinctly concave and not parallel to the other side. The result can be a double image of Jupiter with completely lost bands. A window screen is likely worse even than a glass window because the regularity of the wire spacing will introduce regular errors in the image.

The Moon is often regarded as a rather forgiving astronomical object in the sense that its "wow power" is great with even a pretty ordinary quality telescope.

R.L. Clark, *Amateur Telescope Making in the Internet Age*, Patrick Moore's
Practical Astronomy Series, DOI 10.1007/978-1-4419-6415-1_12,
© Springer Science+Business Media, LLC 2011

Nevertheless the Moon serves as a useful example of just what we are attempting to do in building a quality instrument. Using the 80 mm refractor described elsewhere in this book and a few example values we can hypothesize a focal length of about 30 inches with a 3/4 inch focal length eyepiece. This gives us about 40 power, which is a nice arrangement for that telescope. The objective lens can produce a primary image of the Moon about 1/3 of an inch in diameter. All of the information this telescope is ever going to put in our view of the Moon is going to be compressed into a circle about 1/3 of an inch in diameter.

Suppose we want to look at some detail that is 5 miles in diameter. Even if the desired detail is close to the middle of our view of the Moon it amounts to about 5/2,000 of the image width. $5/2,000 \times 1/3 = 1/120$ of an inch. Everything we hope to see of our detail is jammed into a space of 1/120 of an inch (about 1/5 mm). It doesn't take much loss of resolution to lose details that small.

Common Faults

The Moon is a pretty good object for detecting some common faults. So suppose that your image is pretty sharp in some places and slightly indistinct in other places:

Try rotating the eyepiece while keeping it at the same "in-out" position. If the indistinct spot moves you know that the problem is in the eyepiece. A good possibility is dirt, a fingerprint, dust, or glue on the field lens. The fix is obvious.

If the telescope has a flat mirror fairly close to the eyepiece, such as occurs in a Newtonian reflector or a refractor with a star diagonal, look for the problem there. Dirt, etc., may be the culprit. If the bad area is in the center and is independent of the eyepiece rotation you need to eliminate the eyepiece as a source, so try a different eyepiece. Since you should now be suspecting the flat mirror try shifting it a little if you are dealing with a Newtonian. If you are dealing with a refractor and a star diagonal try shifting the mirror or take the flat mirror out of consideration by testing the telescope as a straight-through telescope.

If you are dealing with a fast (low f ratio) Newtonian and the messy area is in the middle you may be looking at the inevitable result of the central obstruction, which is the diagonal. Perhaps it is bigger than it needs to be. The chapter on Newtonians contains directions and a table for determining the appropriate diagonal size.

A pretty general guide to where to look for the causes of image problems is as follows:

- If a problem manifests itself in some part but not all of the image look for an obstruction at some point somewhere near a focal point. Such a problem might be a fingerprint on the field lens of the eyepiece or a blob of something on the diagonal of a reflector.

- If a problem manifests itself over the entire image or over a symmetric zone of the image look for a problem in the surface condition of the objective or in the alignment of optical components. Incomplete polishing of the surface of a objective will scatter light pretty much uniformly over the image. You get a gray sky with reduced contrast.

Focal Length

Once you have a lens or mirror that you want to use as an objective you will need to know its focal length to a fair degree of accuracy. You probably should not depend on its markings or its description unless you can be sure that the figure refers to focal length measured from the middle, back focal length, or effective focal length.

An effective low-tech way of obtaining the focal length of a mirror or lens is to make a stick with a piece of white cardboard attached at right angles to one end. Stand in the Sun pointing the end without the cardboard at the Sun. You will be able to do that quite accurately by minimizing the shadow of the stick on the cardboard. Then slide the lens up and down the stick until you have the smallest possible solar image on the cardboard. Mark the position of the lens on the stick and repeat the measurement two more times. If you get all three marks pretty close to each other an average should be quite accurate. Reverse the direction of the stick if you are measuring a mirror. But be careful! Don't leave the solar image on the cardboard for more than a few seconds unless you want to start a fire. And don't look at the Sun, particularly through the lens or mirror.

Quality Testing

Many optical components or even complete telescopes offered for sale will have some statement regarding quality. It is important to know what such statements might mean if they are true, how they might be verified, and the extent to which they can tell you anything about the performance of an instrument.

As a first rule, ignore statements about the magnification power of a telescope if they are intended to reflect quality. With the right eyepiece one can get any finite power from any telescope. That does not mean that you will be able to see anything at that power. As a matter of fact it is very rare to have a night in which the atmosphere will support more than 250 power. The Raleigh limit, which indicates that you aren't going to get any more detail than you will get at 50 power per inch of aperture, is mostly correct. Many astronomical objects look best at 20–40 power.

There are at least two rather distinct areas of quality regarding astronomical telescopes:

the quality of the telescope itself as a unit, and the quality of individual components. The two ideas are related in that the telescope can't be any better than its worst part. Since we are concerned with the idea of purchasing some or all parts, then assembling them into a telescope, we need to have ways of estimating the quality of telescope parts.

If, when we use a completed telescope, we are disappointed by what we see we need some sort of trouble shooting advice. Here are a few rough guidelines:

Possible Construction Problems

1. If the problem is lack of definition over the entire field, it may be nothing but a focus problem. Rack it in and out to see if anything changes. If it gets better in one direction or the other you may have calculated the tube length wrong. Check the focal length of the objective or perhaps focus on an object a bit further away. Remember that $1/f = 1/r_1 + 1/r_2$. Only if r_1 is very large do we get $r_2 = f$.
2. If you can reach a "best focus" but it is still not very good, then try looking at a star or an artificial star a little inside, then outside of focus. If the image goes from being elongated in one direction to being elongated in the other direction then you (or the telescope) have a case of astigmatism.

All of our quality considerations are measured against the fundamental question, "How good does it need to be?"

> Have you heard of the wonderful one-hoss shay,
> That was built in such a logical way that
> It ran a hundred years to the day,
> And then of a sudden it – ah, but stay,
> I'll tell you what happened without delay...
>   –Oliver Wendell Holmes

The shay was built in such a way that no part failed until 100 years had passed; then all parts failed and it became "a heap or mound." That is pretty much how we want our telescopes to be. Longevity is not the point. Quality of images is what we are striving for. There is little point to putting a wonderful eyepiece behind a junky objective or mounting a "wonderful one-hoss shay" of a telescope on a shaky tripod that precludes actually seeing much of anything.

Lenses

It is very rare to find a lens advertised as containing some particular type or types of glass. Lenses are evaluated on the basis of their performance rather than their contents. Nevertheless you should have some elementary familiarity with the general ideas associated with lens design.

The amateur telescope maker is unlikely to have any interest in a lens that is not color corrected. A simple uncorrected lens will produce a spectrum or rainbow around star images called "primary color" and which is quite bothersome. Not only do you get the annoying color fringe, but, because the light of different colors is being focused at different places the image suffers and appears blurred.

A color-corrected lens (achromat) is commonly made from two parts – a positive "crown" lens and a negative "flint" lens. The crown component provides the converging power of the lens. Often the converging power of the crown is almost twice that desired of the finished lens. The flint component balances the converging power of the crown back to the desired value. At the same time it reverses the color separation caused by the crown. The reversal amounts to folding the red parts of the image where the crown put the blue parts and the blue parts where the crown put the red parts. The folding is not perfect, so any achromat can be expected to have some "residual color."

Lenses better color corrected than a typical achromat are called apochromats. These are sometimes designed with three or more glass elements and/or using rather exotic glass. In this case exotic means expensive. Either way an apochromat will be a rather expensive item. You should be certain that you will really benefit before you spend the rather considerable extra money.

There is some confusion regarding the term "apochromat." The design often called a Ross Triplet is an improved achromat, but, technically speaking, it is not a true apochromat because it has some residual color error. A true apochromat has no color error. If you are contemplating the purchase of an apochromat be careful to check the manufacturer and the seller.

Mirrors

It is very common for a seller or a manufacturer to make claims regarding the material used for a mirror. The choice of glass type used gives the mirror different characteristics that can make a significant difference depending on the intended use for the telescope.

The common choices are Pyrex, optical plate, and crown. There are also choices regarding the appropriate thickness for a mirror.

Up until the 1960s it was thought that the only acceptable material was Pyrex in a thickness at least 1/6 or even 1/5 of the diameter of the disk. That was prior to anything like the current availability of affordable finished mirrors. Typically the telescope maker bought blanks and invested many hours in the grinding/polishing process. Since an 8-inch mirror might involve a couple of 100 hours of work it made sense to get the best possible raw disk. Pyrex is a little less dense than most other types of glass, making it rather easy to identify. The specific gravity of Pyrex is normally about 2.2, while that of other glass types is 2.5 or more. It is a little faster to grind and polish. The most notable special characteristic of Pyrex is that it is considerably more dimensionally stable when exposed to temperature changes. This means that it is less likely to take on strange shapes and strange optical characteristics when the temperature changes. It also is an advantage in final figuring (polishing) because the shape doesn't change as a result of the heat of friction.

Optical plate glass is commonly used for mirrors and is an excellent material for telescopes. It is also used in much thinner mirrors than would have been acceptable in the 1960s. The difference is probably in the annealing process. Annealing requires that the glass be allowed to cool slowly from its molten state. Often it is done in long stages, during which the glass is allowed to cool for a little, then reheated to almost the previous temperature and allowed to slowly cool again for a little longer than the previous cooling. These cycles are continued for quite a long time until all internal stresses have been relieved. Completely un-annealed glass has very extreme internal stresses. It is subject to shape changes with time and temperature. Ordinary plate glass is likely to provide an inferior mirror, but it is true that any plate glass, or for that matter any glass, will be tempered to some extent.

Crown glass is normally well annealed and is intended for lens work. Because it is used in applications where its transparency is critical it can be more expensive than optical plate. Otherwise it should be regarded about the same as far as use in mirrors is concerned.

Mirror Reflective Coatings

The standard coating for astronomical mirrors is aluminum, electrically deposited in a vacuum chamber. Once coated the mirror should undergo a protective coating procedure. Otherwise the aluminum will be very delicate and subject to oxidization.

You may have been told that aluminum does not rust, or oxidize. It does. This is why your screen door becomes gray almost as soon as it is installed. A nice fact about aluminum oxide (at least for screen doors) is the fact that it seals the aluminum from further oxidation. Previous to the use of aluminum the chemical application of silver as a coating was used. In the hands of a chemist the process worked well, but a failure to control temperature could result in an explosion.

Protected aluminum coatings are quite long lasting. After being damaged or in about 5–10 years, they can be redone. There are numerous contractors who specialize in astronomical mirror coating. They can also remove the previous coating.

Optical Quality Terminology

It is common to describe, or attempt to sound like one is describing, degrees of optical excellence or its lack thereof in terms of angstroms, nanometers, microns, and Lambdas. The use and proliferation of such nonsense doesn't help at all. What people need is something that relates to some measurement that they can actually perceive, for example, the width of a human hair.

Wavelength-Based Measurements

Some folks will attempt to tell you that these are nonsense measurements. They are not nonsense at all, but they need to be correctly qualified. A fairly common claim is something like "1/4 wave accuracy" or "1/10 wave accuracy." These are perfectly valid claims once we agree as to what they mean and where they are measured.

The "wave" part refers to a wavelength of light. Visible light has wavelengths of about 400–700 nm. Yellow sodium light has a wavelength of about 580 nm. Sodium light is the standard in optics, unless something else is stated, and it seldom is. So what is a nanometer? We need to get this expressed in some form that we can relate to real stuff.

One nanometer is 10 to the –9th power of a meter. This amounts to one billionth of a meter. That probably didn't help much. (Incidentally, one angstrom is 1/10 of a nanometer, so 1 ten-billionth of a meter and one micron is 10 nm and one hundred millionth of a meter.) That might be useful if someone tries to confuse you with more esoteric measurements. Otherwise it is pretty much of a "so what?".

One millimeter is one thousandth of a meter, so if we talk about millimeters that gets us down to 10 to the –6th of a millimeter, so we are at one millionth of a millimeter. BUT we have 400–700 of these things, so we can get rid of some zeros and have 4–7 ten thousandths of a millimeter. The sodium light wavelength gives us 5.8 ten thousandths of a millimeter. That quantity is normally called Lambda λ. Converting to inches we get the range of visible light as about 1/64,000 of an inch to 1/36,000 of an inch. The sodium light comes out with a wavelength of about 1/50,000 of an inch. This is still pretty small, but we can relate it to some real stuff. A fairly typical human hair is about 1/4,000 of an inch. So about 200 sodium waves comes out at about a human hair. That, at least, is a "thinkable" dimension.

Rays and Waves

For most, but not quite all, of the thinking we need to do about building a simple telescope we can get along just fine with the "rectilinear propagation of light" principle, which has been around since the ancient Greeks. This principle simply states that light travels in a straight line until it hits something. Rectilinear propagation is the basis of the ray theory approach to the nature of light. It allows us to design things by drawing straight lines with arrowheads to represent where light is going. The straight arrows are presumed to come from the source of the light. We trace them through whatever optical device we are designing. Unfortunately the ray approach, though useful in many situations, is not quite enough to allow us to deal with some quality issues. To discuss some quality considerations we need to have a little familiarity with the wave theory of light.

A point source of light is presumed to be the source of a collection of concentric spheres or wave fronts. An extended source is presumed to be a large collection of point sources very close to each other. For a point source the mental picture of an expanding collection of spheres with a common center is pretty accurate.

Each point of the advancing wave front is considered to be a source in itself. Thus each point is the origin of its own set of "spherical wave frontlets." These wave frontlets cancel each other out in all directions except straight ahead. If we have sodium light the distance between the advancing wave fronts is our friend Lambda λ. It the source of light is pretty far away, for example a star, the spheres become planes. We have an advancing plane wave front. The only reason for choosing sodium light is that it is sort of near the middle of the wavelength range and it is easy to generate. You don't need to burn sodium. Just buy little yellow bulbs from a store. (Radio Shack)

Lens and mirror quality is often represented by statements such as "1/4 wavelength accuracy." We can presume that we are dealing with sodium light, so we have something like bumps not greater than $1/4 \times 1/50,000$ of an inch or not greater than $1/200,000$ of an inch. This is fine, but where are the bumps and what do they do? Suppose we have a step in the surface of a mirror such as in Fig. 12.1. Not that we might actually get such a thing, but let's assume it anyway.

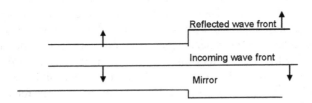

Fig. 12.1 Reflected wave at surface with a step

The wave front hits the mirror, and part of it is reflected as soon as it hits the high area. It then heads back to where it came from. The other part of the wave front continues until it hits the lower area, then it starts back. The reflected wave front has been made to traverse the step twice, once on the way in and once on the way out.

If the step is $1/200,000$ of an inch high it has made an adjustment in the wave front of twice that, or $1/100,000$ of an inch. A 1/4 wave bump in the mirror will make a 1/2 wave bump in the wave front. It is the wave front that we actually see. When a seller claims 1/4 wave accuracy or 1/10 wave accuracy you need to know if he or she is talking about accuracy on the wave front or on the surface. In most cases you will be told. If you aren't, then the seller doesn't really know what they are selling. That doesn't necessarily mean that you can't use the item they are selling. The "good enough" rule applies. If 1/2 wave is good enough, then their 1/4 wave claim is OK if it is true in either way.

There is yet another problem in the wave measurement of accuracy. You will see claims about RMS error. RMS stands for "root mean square." RMS is a way of treating errors and is a standard technique in many fields of engineering. The process is to take a list of values, say 1,3,2,9,1, and 2. If you averaged these you would get 3.

If we add up the squares of 1,9,4,81,1, and 4 we get 100 as the sum of the squares. If we average those guys we get 16.666. Then we take the square root of that and get about 4.1. So the RMS value is the square root of the mean (average) of the squares. If that reminds you of the standard deviation function in statistics, it should.

The RMS function can be applied to either the surface variations or the variations in the resulting wave, so now we have four variations.

It may well be that an optical item that you find available via the Internet was purchased as surplus by the person/organization currently offering it for sale. The seller may only be able to tell you what the original seller told him or her. An item, or its packaging, may be marked "1/4 wave" or some such measurement. If you ask if that applies to the surface or the wave front the seller may have absolutely no way of knowing.

There are a couple of variations still left to discuss. You surely noticed that the mirror defect (a step) depicted in Fig. 12.1 is very artificial. You are simply not going to find an optical surface with a step (unless it has been broken and glued back together, which is pretty unlikely).

Let's assume an optical surface (a reflective surface, although a refractive surface works about the same way) is, say, a 24-inch circle. Suppose it had, relative to the desired shape, a smooth hill in the center so that the center was 1/4 of a wave high and the hill extended evenly to the edges of the mirror. If the surface was concave, such as a mirror for a 24-inch Newtonian, the defect would be pretty easy to ignore. The focal length would be just a little longer than intended. Rack the eyepiece out a very little and all would be fine.

If the same surface had many little hills, each 1/4 wave high, the performance of the mirror might be rather poor. The image would be quite confused. The reflected rays would go all over and certainly would not form a decent image.

As Fig. 12.2 illustrates, it is not so much the height of the defect as it is the slope implied by the defect that causes damage to the image. Some professionals describe a surface with no more than 4 angstroms of surface roughness as a "well-polished optic." That is normally sufficient for most telescopic uses.

In general, specifications like "1/4 wave accuracy" refer to the entire surface of the item. Large items may be described by statements such as "1/4 wave each 20 mm." The 20 mm refers to a 20 mm circle. The meaning is obvious.

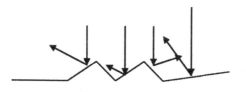

Fig. 12.2 Reflected rays at a rough surface

Scratch and Dig Optical Surface Specifications

These come from a US military specification (MIL-STD) and measure the maximum scratch or dig on an optical surface.

An item may be listed as 60–40 scratch and dig. The first number refers to scratches and the second to digs. A dig is, typically, a small (or not so small) pit left by the grinding operation that was not removed by the polishing operation. The numbers are expressed in tenths of microns or nanometers. A 60–40 surface may have scratches up to 60 nm wide and digs (pits) up to 40 nm wide. This is, respectively, 0.00024 and 0.00016 inches wide. A 60–40 surface is reasonably good. The effect of scratches and digs is to scatter light. The result is a gray sky where it should be black and a loss of detail.

Scatter (Spot) Diagrams

A relatively recent measuring tool is the scatter diagram. Its importance is to show optical system designers where the light entering an optical system actually ends up.

Strehl Ratio

The Strehl ratio is a relatively new measure and is used by some manufacturers. The mere fact that a seller/manufacturer quotes the Strehl ratio probably indicates some good things. The Strehl ratio can be thought of as the decimal fraction of the light that lands in the correct place in an image to the amount of light that "belongs" in an image. To get a little further in understanding we need to do a little more with wave theory:

Even if our optical element were absolutely perfect, not all the light from a star would end up in a perfect mathematical point. Since a mathematical point has no size, how would we see it and how would we think about the infinite concentration of energy that would result? What happens, even in the theoretically perfect case, is that we get the light distributed in a sort of target pattern (Fig. 12.3).

The central blob is called the Airy disk. The rings occur because the light travels different distances. When the difference in distance is exactly half a wavelength it cancels itself. It happens that the diameter of the Airy disk decreases as the aperture of the lens or mirror creating it increases. It is entirely independent of the focal length of the objective.

The Airy disk should contain about 84% of the light from a star. The first ring should contain about 7%, and the second should contain about 3%, etc. The Strehl ratio compares the light going into the Airy disk with the light that should go into it. A value of 1.00 would indicate perfection – all the light that theoretically should be there is there. A value of 0.5 indicates an image that is at best terrible. Only half the light that should be there is there. Light that is missing from the Airy disk must go somewhere. The effect is, essentially, doubled. A value of 0.82 is roughly equivalent

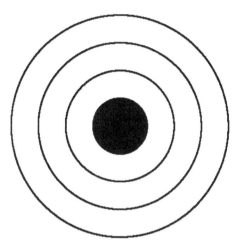

Fig. 12.3 Diffraction pattern

to the Raleigh limit. A value of 0.94 can be called excellent. For a more complete discussion of the Strehl value with relationships to the other, older criteria, go to web page http://www.rfroyce.com/standards.htm.

It is likely that you have heard of the Raleigh criterion or Raleigh limit. We discussed it in a earlier chapter. This value is often used as both a definition of satisfactory optics and as a "theoretical" resolution limit beyond which it is impossible to go. As a resolution limit the Raleigh criterion is often stated in the following form:

$$\text{Maximum magnification} = 50 \times \text{Objective aperture in inches}$$

The idea is that, although you can increase the magnification of a 2-inch telescope arbitrarily, any magnification above 100 will show a bigger image but will show no more detail. What this rather arbitrary statement comes from is the observation that if the Airy disks of two stars overlap by more than about one third we are unlikely to discern them as individual points. This is critical for observers interested in splitting double stars. If it is used that way it is a rough but acceptable measure.

Here is a list of well-known doubles with their angular separation. Your friends may be suitably impressed by what you (or, more properly, your telescope) can split:

Mizar and Alcor	Naked eye
Mizar	2-, 3-, or 4-inch should be easy; separation is 14.4 inches
Epsilon Lyra	Naked eye double, quadruple for 4-inch; separations are 2.6 and 2.4 inches. Not an easy target for a 4-inch. You may get two elongated images rather than distinct separation

Quality of "Flat" Mirrors

Flat mirrors are used as diagonals in Newtonian telescopes and star diagonals for refractors. They are also used whenever we want to "fold an optical system." The obvious question is "How flat is flat enough?" The answer depends on what you are using it for and where it is in the optical system. Following are two extreme examples.

Plate glass store windows are certainly not designed to be optical instruments. They often have very severe bows and bulges. If you stand close to one the reflection will be reasonably good. It will certainly suffice for you to comb your hair. If you look at your reflection from the other side of the street, it will be pretty bad. Welcome to the fun house. The glass (mirror) is the same. The difference is where it is in the optical train. If the reflected image has to travel only a short distance, a pretty poor mirror is still OK. If it has to travel a long distance, then the effect of a bow or bulge is exaggerated. The geometry should be pretty obvious.

If you use a piece of regular window glass as a mirror it will work OK as long as you are using it fairly close and more or less square on. If you attempt to use it at a very shallow angle so the image just "glances" off the surface, the results will be very bad. This is more or less the same effect as observed with the store window across the street. The mirror used as a diagonal in a Newtonian or in a normal star diagonal operates at a 45-degree angle. Some star diagonals are designed to operate at a 45-degree angle; accordingly, the flat mirror in such a star diagonal operates at a 22.5 degree angle. Thus the mirror in such a diagonal needs to be better (flatter) than one in a standard star diagonal.

Testing a Flat

Norman Remer's book *Making a Refractor* has an entire chapter on making a flat. Although you probably don't want to make it, you do want to make an intelligent purchase and be able to determine its quality. So, for your purposes the important part of that chapter is the material on testing a flat. Remer provides plans for a nice interference tester that works very well. It produces results that are pretty easy to interrupt and provides a nice demonstration for those of you curious about what optical precision is all about.

The principle of an interference tester is to place two pieces of glass with reflective surfaces very close together and almost parallel. One resting on the other with a small piece of tissue between them at one edge works well. One of the surfaces should be known to be flat. The other is the surface that you are testing. The two surfaces enclose a very thin air wedge. At different spots the wedge has different thicknesses. A monochromatic light is shone from above.

As the difference in separation of the two surfaces goes from being an odd multiple of the wavelength of light to an even multiple (making it an integer number of wavelengths), the individual wave fronts alternately cancel and reinforce each other. It is important that the light be pretty close to monochromatic. Viewed from above (same direction as the light source) looking through the top one sees a pattern of dark and light lines. If the lines are straight the tested surface is flat. If the lines bend then the degree of variation (in wavelengths) of the test surface from flat can be measured by the "number of fringes of bend."

Put the pieces together very carefully. Don't slide one on the other. It is necessary that the top piece be reasonably close to clear. A fine ground surface will work. The bottom surface can be aluminized. That may make the fringes a little harder to detect, but they will be there.

Obviously testing an unknown surface requires a known flat. Try to use a flat belonging to a local college, university, or astronomy club. Bring your unknown piece to the flat and try to get the owner of the flat to help you with the test. People who make flats of very great precision have many hours invested and are likely to be very protective of the surface in which they have invested a lot of time.

It is possible to use a water surface in a bowl as a reference flat. Such a surface will take on the same curvature as the surface of Earth. That is close enough, but you have to avoid air currents, heavy-footed ants, and furnace turn-on cycles.

For a Newtonian diagonal or a refractor star diagonal a mirror is probably flat enough if you can find a piece which is flat, at the surface, to 1/4 wave. That gives you 1/2 wave accuracy at the wave front.

If you want to use the flat as a relay in an instrument such as a folded Herschelian telescope you will need to get a better flat. The reason is the greater distance between the relay mirror and the image. If the relay has some, perhaps very small, areas not parallel with the rest of the mirror the light reflected from those areas will go off at a slightly different angle from the light at the remainder of the mirror. It doesn't take much of a divergence in angle to cause the light to end up outside the Airy disk.

The flats available on eBay, etc., without much in the way of dependable specifications, may have decent areas. You can also pull flat mirrors out of discarded copy machines. Those flats are commonly about 5/8 inch wide and 8 or 10 inches long. Cut pieces with an ordinary glasscutter, then, test each piece. Cut first, then test, because cutting may release internal strains (or remove them). See Remer's book for more details.

There is an excellent article by Daniel W. Rickey in *Sky & Telescope*'s June 2006 issue. It examines the question of optical quality from a qualitative point of view and provides examples of variously degraded images. The examples are particularly useful, because they provide "palpable" comparison. Planetary images are presented to illustrate just how an image can be expected to look under various conditions. You can't ask for anything more direct and applicable than that.

Optical Aberrations

This section discusses optical aberrations from the point of view of the amateur telescope maker. No attempt is made to provide mathematical definitions or to do any sort of complete description. There are many books in the areas of optical engineering that cover that material.

The optical aberrations that may, or probably will, to some extent, show up in your homemade telescope are:

- spherical aberration
- coma
- astigmatism
- curvature of field
- distortion
- chromatic aberration

These same problems may, and probably will, occur in a commercial telescope. In all cases the terms describe situations in which light from the object being observed is ending up someplace other than where it belongs. Other problems can come from defects in the manufacturing process. It is difficult to differentiate between some mechanical problems that can come from manufacturing and design problems that are more deeply ingrained. For example, "turned down edge," a manufacturing defect so common that it has a very specific abbreviation, TDE, can appear as under correction for spherical abbreviation.

Other manufacturing problems such as incomplete polish are quite different from the standard aberrations.

Spherical abbreviation is a characteristic of a spherical lens or mirror surface. Rays parallel to the optical axis of the surface and very close to the axis (thus close to the center of the lens or mirror) will come to a focus at a slightly different distance from the surface than rays closer to the edge.

This aberration depends on the diameter of the lens or mirror. More specifically it depends on the cube of the f ratio. For that reason high f ratio systems can be essentially free of spherical abbreviation. Faster (lower f ratio) systems can be "figured," that is, altered from perfectly spherical form, to reduce or eliminate the effect of spherical abbreviation.

Typically, a Newtonian reflector must be made to be a parabola if it is designed to f8 or faster. With slower mirrors the difference between the appropriate parabola and a sphere is so small that it has no real effect. The optical test most often used to determine if the surface of a mirror is the correct parabola is called the Foucault test. The Foucault test is often considered rather mysterious but is really rather straightforward, if you are willing to devote a couple of hours to understanding it. There is an excellent presentation on the process of making a spherical mirror into the appropriate parabola presented by the web page http://bobmay.astronomy.net/foucault/harbor1.htm. Also Texereau, in *How to Build a Telescope*, presents an excellent plan for building the appropriate test device.

A star image should appear as a round and very small dot. You are not going to get a true point image, but it should look that way. After all, a true point has no size. If we actually did get true points as images we wouldn't be able to see them.

Astigmatism, as it affects the human eye, was discussed earlier. In an optical instrument astigmatism is, essentially, the same as in the human eye. It may be caused by a defect in the shaping of some lens or mirror such that it is slightly cylindrical or, much more likely, by a failure to mount some lens or mirror square to the optical axis. In a telescope astigmatism shows up by causing the image of a star to be elongated in one direction when the eyepiece is inside of focus and at right angles when the eyepiece is outside of focus. In extreme cases the image becomes a line in one direction and then a line in the other direction.

One might think that the idea that if the image is distorted one way inside of focus and the other way outside of focus it doesn't matter, since we only really care about the image at focus; we don't really need to worry about a small amount of astigmatism. Not so! The position that we are calling "at focus" is really just the "best focus" between the two distortions. An astigmatic instrument, when operating at its best focus, is enlarging the point image in both directions. A good description is to say that it is producing a "soft" image. We don't want soft images. We want hard, crisp images. A soft image spreads the light from a star out enough to make very faint stars invisible! The image of an extended object, for example the Moon or a planet, may look pretty good, but softness of the image will cause you to lose fine details. A pretty good test is to examine the floor of the Moon crater Plato. Look for the "craterlets" in an otherwise flat-looking surface. At 100 power can you distinguish the rims and shadows?

An eyepiece can be corrected to accommodate the user's astigmatism. Such a design amounts to imposing an equal and opposite astigmatism on the telescope. That process works well but produces an eyepiece that has to be marked regarding "which side is up" and which can be used only by the person for whom it was built.

If you would like to experiment with astigmatism correction one method is to add to the eyepiece a weak positive or negative lens that is intentionally mounted at a very slight slant. Another approach is to get your optician to supply a lens that matches your prescription so far as the astigmatism is concerned but which is of zero power. That is, it ignores your nearsightedness or farsightedness. Using the same machine that fits lenses to frames the optician can trim the lens to fit in the eyepiece. Probably the best thing to do is to slip this into the eyepiece tube on the telescope side with the convex face pointing outward, toward the telescope. Take special care to keep the surfaces clean, because you will be installing it very close to the eyepiece's focal point. Fingerprints will look like ruts in a muddy road and dust particles will take on the appearance of boulders. Ask the optician to mark "up" somewhere on the edge, and make sure you get the lens in square. Otherwise you will be adding more astigmatism, probably in a different direction.

Of all the problems that can beset either a homemade telescope or a commercial instrument, astigmatism is the most common. In most cases the astigmatism is caused not by a defect in any optical component but by a defect in the alignment between components. Because it can be corrected by essentially mechanical improvements or adjustments it is much easier to fix than the problems inherent in the actual optical parts.

Quick and Easy Tests

These are quick and easy tests that can be applied to an optical element (not a complete telescope). Because they are quick and easy you cannot expect them to provide more than a good guess as to the suitability of something for use in a telescope. These tests are pretty much limited to what you can do while standing in a store.

1. Look at the specs. If the specs don't indicate a chance of utility then you can hardly expect it. For example: Is the focal length suitable for the intended use? A lens with a 10-inch focal length is not going to serve as a part of an eyepiece.
2. Look at the surface. A few scratches are probably not going to hurt, but a lot of pits will. If the item is a mirror, look at the condition of the coating. Does it show signs of rough cleaning? Does it look corroded? Does light show through it? Any light getting through is not going to become part of an image. It may well need re-coating and/or re-polishing.
3. For potential objective lenses try them out with an eyepiece of known quality. Rig some way of keeping the objective and the eyepiece lined up and square. A cardboard tube of an appropriate length and having square ends will work. The game of holding an objective in one hand and an eyepiece in the other to test either is nonsense.
4. Do you get a lot of extra light (flare) outside the image when you look at a distant light such as a streetlight? The use of a Christmas ornament as a point source is pretty common. Do you get about the same set of rings inside and outside of focus? Do you get distortion? For eyepieces, try them out with an objective of known quality.
5. For surfaces that are supposed to be flat, look at a reflection of something regular such as a brick wall at a very shallow angle – as much of a glancing angle as you can manage. Are the lines straight? You don't want something that reflects like a pond surface with ripples.

Chapter 13

Junk Collecting and Working with Modern Materials

Junk Collection (Creative Cross Utilization)

A characteristic of amateur telescope makers and other such groups is pride in utilizing mundane materials and items in their exotic creations. To make it sound properly esoteric, let's call it "Creative Cross Utilization." That justifies, hopefully, maintaining an inventory of mundane materials and items – in other words, junk collecting.

There are two fundamental approaches:

1. Search for something when you need it.
   Search through someone else's collection of stuff. For example, once you have a general familiarity with PVC tubes and fixtures you can walk the aisles of the local plumbing supply store and likely come up with the correct combination for some specific need.
2. Acquire stuff just because it might be useful.
   Useful someday, for something. This approach is the most serious form of the affliction. At least, as telescope makers, we don't have the same kinds of collections as do the auto fanatics. We are unlikely to end up with a yard filled with junk cars. Or are we? An engine block can make a very steady mounting. See Russell Porter's illustration in Volume I of *ATM*. It looks like an old Buick straight eight. This kind of Creative Cross Utilization can result in a filled basement and/or garage.

A suggestion regarding salvaging machine parts. Often, when salvaging a machine for parts, it is useful to leave some subassemblies together. Leave bearings,

R.L. Clark, *Amateur Telescope Making in the Internet Age*, Patrick Moore's
Practical Astronomy Series, DOI 10.1007/978-1-4419-6415-1_13,
© Springer Science+Business Media, LLC 2011

shafts, and gears assembled if you can. If you need to use a certain shaft you probably need the bearings that fit it.

Unfortunately, most of us are hampered by the desires of spouses, parents, or other relatives who feel that their personal reputations are injured by association with junkyards. It has also been rumored that certain landlords have objected to accumulations of such treasure.

Here, now, are some specific ideas about what you can usefully collect:

- Steel rods from dot matrix printers are, unless damaged, very straight and well polished. They are ideal for Foucault machines or related devices (anything where you need sliders – rails – for precise longitudinal positioning).
- Any optical device is probably worth saving. Small camera lenses and even small projection lenses often make good eyepieces. With the advent of digital photography for the masses, old, good quality 35 mm film cameras are often available for very little at yard sales.
- Xerox machines (lens and 1st surface mirrors).
- Projectors for their lenses, condensers, etc.
- Large kitchen mixers can become lens/mirror grinding machines.
- A 4-inch chrome ring from some kitchen appliance with a rim about 1/4 of an inch wide becomes a retaining ring for a long focal length, 4-inch mirror with a defective edge. This makes the mirror entirely usable as the objective for a long-folded Herschelian telescope.
- Some cans, for example those for nuts and coffee, have seals consisting of thick foil that is peeled back to open the can. What is left is a neat metal ring with sides and a lip. If the main part of the can is cardboard, the ring can be removed with ease. If the main part is metal, it can be cut away from the ring, sometimes with a can opener. The rings can then become lens cells or mirror cells. Often the sizes are ideal, 3- or 4-inch, and they fit around lenses or mirrors just right.
- Computer printers normally contain several stepper motors. We don't discuss powered telescope drives in this book, but in the event that you want to build one a stepper motor is a good place to start.
- Timer motors from washing machines, driers, etc., are often gear-head motors already reduced to convenient speeds such as 1 or 3 rpm, etc. These can make a good beginning for a "clock drive." They can also be obtained from numerous salvage companies online.
- Weights from weight lifting sets can be easily used as counterweights for a variety of mounts. They come in many sizes and have a hole through the center that makes attaching them easy.
- If you don't have any neighbors who have decided that weight lifting as a career just isn't going to work you can make a weight by filling a can with concrete. Concrete weighs about 145 lbs per cubic foot. Select the appropriate can (coffee, etc.) and poke a hole in the center of the bottom big enough to put a tube or small pipe through. Set the can upright with the tube sticking out a little and up through the center of the can, pour it full of concrete and let it set up. You don't need to get a very good fit around the tube through the bottom of the can. Wet concrete doesn't do much flowing.

Lens Restoration/Used Lenses

Once your friends and relatives get the idea that you are messing around with optical stuff you are likely to receive a lot of items that they might have (perhaps should have) thrown out. See the section on "creative utilization."

One of the most interesting versions of optical junk is that of old or antique cameras and/or lenses. Sometimes you will find, for sale, antique lenses with various "fatal" defects. Sometimes they are given away by folks who have no clue as to what they are or what they might become. They may be available very cheaply. Sometimes the defects aren't fatal at all. Lenses from high altitude cameras are very nice. Old lenses with designs called "Rapid Rectilinear" or builders such as Dalmeyer may be valuable antiques and/or good objective lenses. Often all that is needed is a bit of cleaning and a new tube for a telescope or a new body for a camera.

If a lens has serious scratches there isn't much that you can do. If you decide to grind or polish it out you will probably disturb the curves enough to make the image much worse than it was. To do that job correctly you need to, for all practical purposes, re-grind the lens almost from scratch. In some cases lenses can be re-polished if they have very fine scratches or pits. See the section on polishing ideas later in this chapter. The need to polish is more likely to occur with a used mirror, which may not have been completely polished when the mirror was first made, or you may want to improve the "figure" of a mirror that was never brought to a good parabolic surface.

If a lens is in need of anything that can be done it probably has problems with its cementing.

Lens Cementing

Lens cementing is commonly needed on old lenses. The appearance of a lens with deteriorated cement is bad enough to make most people think that the situation is beyond correction. It is not. You just have to know what to do and to do it carefully.

Some high quality lenses are what are called "cemented doublets"; typically these lenses have two glass elements that are separated by some material. The appropriate "cement" for older lenses is balsam. Balsam is a product made from purified pine pitch. It was chosen because it has optical properties very close to those of "crown" glass and is very clear (for about 50–75 years).

Modern cemented lenses use a material that is hardened by ultraviolet light. Some amateurs use ordinary mineral oil and apply the elastic type of electrical tape around the edges to keep the oil from evaporating. Believe it or not that method produces a neat job, and the oil stays where it belongs. In addition the oil does not have any negative effect such as dissolving the tape and migrating color between the lenses.

Getting a Lens Apart

You may have the opportunity to acquire an oldie but goodie lens with damaged cement. As it ages the balsam may turn orange or brown or black and may craze. As you would suspect, this junk is very ugly and has a very serious negative effect on the quality of the image. You need to remove it and replace it with something.

A lens that was designed to be cemented will work if it is air-spaced. Once you get the thing apart make a thin paper donut for the outer edge and carefully put it back together with the paper acting as a separator between the two pieces. The main problem is that you have taken a lens with two air-to-glass interfaces and made it into one with four such interfaces. The result is that less light will get transmitted; also you may get reflections from the newly created glass-air interfaces.

The first, and non-trivial, step is to get them apart without damage. There are several web page treatments of this process:

> www.skgrimes.com/popsci/index.htm describes taking them apart by carefully heating and re-cementing with a UV-activated cement. (Look for it on eBay, etc.)
>
> wapedia.mobi/en/Talk:Achromatic_lens
>
> and
>
> www.rangefinderforum.com

A web search on "balsam" or "cemented lens" will yield several reasonable approaches to the process of removing old balsam and replacing it with new cement. The word "reasonable" does not mean that they will always work or that they agree with each other. What follows is a combined method that has worked and minimizes the chance of damaging the lens. It was applied to several Dolmeyer lenses about 150 years old. It yielded lenses as just good as they were when new, and that is pretty good.

This process uses lacquer solvent that is extremely flammable. Its vapor is both toxic and explosive. Avoid open flame, including pilot lights, etc. If you are careless you can relocate yourself and the remains of walls and kitchen appliances to the great outdoors.

1. Carefully remove the lens from its cell. This may involve lots of WD40 and some time. Be careful of the cell threads. They are very fine and can easily be assembled cross threaded. You may need to apply a little "drift" to get threaded rings to move after 100 years or so of no motion. Cut a very small V-shaped slot in the edge of a stuck ring. Put the tip of a sharp cold chisel in the slot and tap it in the appropriate direction to get it to loosen. Even a very gentle tap may get motion from stuck joint. Once it has started the WD40 will work its way in. There may be a lead-foil ribbon around the edges of the cemented pair. Remove it carefully. A sharp knife may help.

2. Clean the lens as it is, still cemented together. The idea is to remove any bits of grit that might cause later problems.

3. Place it in a saucepan covered with water and heat slowly so you don't put any strains on it. Once you have it at a low boil fish it out and put it on a clean cloth. Note: The heating may have done the job. Several of the experts claim that it will have. Try getting the parts apart while they are still fairly hot. Use a soft cloth to hold them. If the two glass components don't come apart the cement will have, at least, softened so the danger of temperature induced cracking is reduced and the next step will work easily.

4. While the pair is still good and warm to the touch place them into a coffee can with about 1 inch of lacquer thinner. Cover the can so you won't have lacquer solvent fumes floating around waiting to explode. Let it set outside, at least overnight. The warmth of the glass seems to draw the solvent in between the pieces. As soon as a little gets between the lenses more will follow and you have it made, or, in this case, unmade. The two parts should come apart easily.

5. Clean the remaining junk off the lens parts with the solvent and safely get rid of the remaining solvent.

6. Put the two glass parts back together according to the directions supplied with whatever cement you have purchased. Some you can use right out of the bottle, some you will have to dissolve, again using lacquer solvent. The general idea is to put a drop on the clean concave surface and set the clean convex surface onto it. The drop of cement should flow out to the entire surface without bubbles. If a little flows out the edges no harm is done. You can clean up later. If the lens is larger than 3 or 4 inches consider using the mineral oil technique so that temperature changes won't set up one piece of glass stressing the other. If the lens never had anything between the surfaces leave it that way.

Polishing and Coating Ideas

This discussion is, in part, intended to help you to evaluate a lens or mirror made by someone else as well as fill in some holes in other discussions if you chose to grind and polish your own optical surfaces. If you have acquired a mirror with a less than satisfactory figure it makes sense to work it to a better parabola. There are several books and numerous web pages that can guide you through the polish – then test – then polish – etc. process involved in figuring a mirror. Texerau offers one of the best descriptions. If the mirror has a dulled surface coating resulting probably from age, all that you are likely to need to do is get it re-coated. The organizations listed in the "Sources" appendix can remove an old coating prior to their applying a new one. An aluminized surface more than 10 years old will probably need re-coating.

If the purpose is to improve a used mirror you will probably have to remove an existing aluminum coating. There is a chemical sold in electronic hobby stores for cleaning solder off circuit boards. It is said to do a good job of removing aluminum coatings on mirrors. It may slightly etch the glass surface, so don't use it unless you expect to re-polish the surface. If all you need is a re-coating let the coating contractor do the job.

It is assumed that, if you have embarked on a grinding/polishing expedition you have progressed beyond the first stage of telescope making. It is thus assumed that you have read some one or several of the books that cover the making of your own optical parts. What follows is not a full set of instructions. See Remer and/or Texerau for that. This is simply a collection of suggestions that may serve to clarify or reinforce what is provided by those references.

No Drag, No Progress

A small amount of resistance indicates that something is actually happening. The drag may be mostly at the beginning and/or end of a stroke.

Make your rouge-water mixture more toward the thin side than toward the thick side. After an hour or so the pitch surface will be pretty well infused with rouge. It is the rouge embedded in the pitch that really does the polishing.

How Long Will It Take?

There is no getting around the fact that polishing will require at least as much time and work as did all of the grinding. Ten hours with rouge is an absolute minimum. Up to 15 h is likely to produce an OK surface, while 15 or 20 h is even better.

Polishing should proceed from the center outward. The outside ring of a mirror or lens will be the last to come to a polish. If you use a strong light reflected from the surface and a strong magnifying glass to examine the surface of a commercial lens or mirror you are very likely to find many pits in the outer 1/4 inch or so. This is not being pointed out to justify incomplete polishing but to provide proof that, if you choose, you can do better than commercial grade work. An incomplete polish will scatter some light. Since scattered light goes somewhere it can be expected to land in places that are otherwise black, since most of a field of view is black. It will reduce the contrast between stars and their background. An indicator of a poorly polished surface is a gray background sky. The likelihood that the margins of a lens will have residual pits is *one* of the reasons why you may find that masking out the edge will improve the performance.

There are several differing opinions on the length of a session of polishing. Some authors suggest that long sessions are best and that a session of less than 30 min is wasted or worse. Others express a concern about the shape of the lap changing as a result of more than 30 min continuous polishing. Several sessions of 30 min each per day seems to be a reasonable compromise that gets the job done.

The first 30 min session, or even 15-min session, will produce a noticeable pseudo-polish. If the surface is intended to become a mirror some testing can begin at that point. As polishing continues apparent progress slows. Up to the 6-hours point examination of the surface every hour or so will show progress. After that don't expect to see much change any more frequently than after 2 hours.

It is sometimes difficult to tell just what is going on at the edge, say the outside 1/4 inch. Examine the surface with the naked eye or with a magnifying glass. If you are working on a mirror examine by reflection, at an oblique angle, with a strong light. Hold the work so that the edge that you are examining is away from you. If you hold it so that you are examining the edge closest to the light you may see the effect of the chamfer at the edge or even the edge itself because you will be seeing through the glass. Try various magnifying devices. A low power microscope may work well for you. If you own a laser pointer shine it on the surface. This is a very demanding test. If a *very clean* surface disappears in the light of a laser it is pretty free of pits. (Exercise care, though. You don't want to hit yourself in the eye with a laser.)

Another way is to try holding the piece about 2 feet from a fairly strong light source such as a 100 W bulb. Let the light come through the work, and examine the surface with a strong magnifier, say a 1 inch or less focal length lens. The idea of the short focal length is to provide such a small depth of focus that you can be sure which surface you are looking at. You can call it a good surface when the pits at the edge can qualify as "isolated."

If a mirror (more likely the edges of a mirror) is not polished enough it may become apparent after the reflective coating is on by it looking slightly "frosted."

It is, obviously, very important to protect both the lap and the work from anything that can produce a scratch:

- Don't have grinding go on in the same room, at the same time, as the polishing.
- Don't go from grinding to polishing without a change of clothing.
- While pressing or just resting cover both the work and the lap to protect them from dust and dirt falling from the ceiling. An inverted coffee can or bucket will do the job.
- When pressing the top item (either lap or work) can slide on the lower item, particularly if they are lubricated with soap. The result can be a bad pressing job if the sliding stops with an overlap creating a ridge in the pitch. If it doesn't stop something can land on the floor. This is probably a worse version of a pressing job.

If you do get a scratch during polishing don't go into a state of complete despair. It may polish out. Even if it doesn't polish out a scratch will have much less effect on the image than a bunch of fine pits. If you get a chip you can neutralize that entire part of the surface with a little black paint.

It is important to realize that a serious scratch or even a chip is not as detrimental to optical performance as it is to cosmetic appearances. Think about it this way:

The chip or scratch renders some part of the surface unable to make a positive contribution to the performance. This is the same, but no worse than, the effect of having a smaller surface. It would take a pretty massive chip to cut a 6-inch mirror down to the equivalent of a 5 1/2-inch mirror. It would require a round chip with a radius of 1.198 inches.

The chip or scratch may scatter some light into the image. This is much more serious than the light that is blocked because it reduces contrast. Contrast is what you need when you are searching for a nebula or galaxy. You can cure the scattered light problem with a little black paint. A lens or mirror may become available because its owner thinks it is damaged beyond repair. All the better for the buyer.

A crack is an entirely different matter. Without any doubt the surface on each side of the crack will lose its orientation with the other. A mirror with a crack down the middle would work better if you simply discarded half of the mirror.

Lap Making

There are two distinct schools of thought regarding how to make a lap with channels:

1. Make up individual squares of pitch. (Texereau)
2. Impress the channels into a complete lap that has not yet completely cooled. (*ATM* Vol I)

Use whichever seems easier to you!!!!! A lap does not need to be pretty, and, in fact, after a few hours of use it will be thoroughly ugly.

Either way you must have distinct channels unless the lap is a very small one, like 2 inches. The channels provide space for expansion of the faucets. Otherwise it will be impossible to maintain good contact between the lap and the work.

If you are making a lap by pouring molten pitch onto a form keep in mind the fact that the pouring process will begin the cooling process. Get the pitch a little hotter (thus more liquid) than it seems to need to be so it will spread fairly evenly. If you get a lump due to failure to flow well enough don't try to press it out. Re-melt and pour again. Lumps just don't seem to press out.

Pitch is described as hard or soft. That is a qualitative measure applied after it has cooled. If it is too hard it will be difficult to cut channels into it, and it will scatter a lot of chips. If it is too soft it will easily take the impression of a thumbnail after it has cooled. To soften pitch add bees wax and let it melt (5% of the volume is likely to be enough). To harden pitch just cook it (keep it molten) for a few minutes. Get bees wax by buying a pure bees wax candle.

There is no harm in re-using pitch as long as it hasn't been contaminated by sand or grinding grit. The accumulation of rouge does not seem to do any harm. Re-use with the associated melting is the typical reason for needing to add bees wax.

A lap that is slightly smaller than the work piece is OK. It will somewhat protect you from making a turned edge. Bigger is not OK. It is very likely to turn the edge.

Use a fair amount of liquid soap added to the rouge between the surfaces or directly squirted on the surface during the initial cold pressing. A fresh lap is more likely to grab or stick to the work than is an older, broken-in, lap. The soap will serve as a lubricant early in the process when you are sliding the surfaces around to get the pitch to conform to the work. The same easy slide is not desirable once you have good contact and need some drag to get some polishing done. Rinse the soap off of both surfaces once you have good contact.

If the work piece and the pitch lap get stuck together it is probably the result of the interface drying during pressing. Perhaps the pieces were left to cold press overnight but that turned out to be two or three nights.

You can always melt the pitch by putting the whole thing in a pot of water and heating it on the stove. There is no danger that the sticking of the work to a pitch lap will result in destruction of the work piece (lens or mirror). If you have to separate them with heat you will, of course, need to make a new lap.

Like most problems it is easier to avoid sticking in the first place than it is to fix it. If you cover the pair with a wet cloth when leaving them for pressing you may avoid problems. Suppose they did stick. Try placing the pair in a bowl of water deep enough to get the interface wet. After an hour or two, water will penetrate the interface and two are likely to slide apart.

Since this is a safe approach try it even overnight before you attempt more drastic approaches. No harm can be done by soaking.

Here's a brute force approach to the problem. The old faithful Jorgensen clamp may work. There is some danger of breaking off significant parts of the lap. (Small parts missing, particularly at the edges, won't do any harm.) A Jorgensen clamp is a tool sometimes called a hand clamp. It is made from two pieces of wood and two threaded steel rods. To use it to separate disks put one jaw against the edge of one of the disks and the other jaw against the edge of the other disk. Tighten the clamp to make it exert a sliding force between the two disks. Be patient. Pull up moderately and wait a few seconds. If nothing moves pull it up a little tighter and wait again. Eventually one will slide against the other. After a little motion has occurred it is important that you keep it going so the disks don't freeze again. Work over some thick towels or carpet. The disks may come apart quickly.

Working with PVC Pipe

PVC (polyvinyl chloride) pipe is relatively new to the amateur telescope makers' collection of materials. It has several plusses and only a few minuses. One of its problems is the fact that it uses diethylhexyl phthalate as a plasticizer. The Food and Drug Administration has advised caution in handling it, but there seems to be little real danger for using the way we are likely to use it. If you are sawing, grinding, or sanding it, wear a mask to avoid inhaling the stuff.

PVC pipe is very good as a telescope tube and can be worked into several versions of focusers. The various connectors, caps, etc., can be fashioned into lens cells and a wide variety of additional telescope gadgets such as mounting piers (see Chap. 11).

A wide variety of sizes and shapes are available at hardware stores. The pipe comes in both a thin wall and a thick wall form. The thick wall variety is probably the best choice for telescope tubes. Hardware stores seem to carry the pipe with inside diameters up to 6 inches. Sizes up to 24 inches are manufactured. The cement sold for the stuff is very good. It works by actually welding the parts. It is important to have good contact for a joint. The fumes from the glue are very toxic. Avoid inhaling them. Likewise the gas generated when PVC burns is toxic. Don't burn the stuff!

PVC can be sawn, either by a power saw or by hand. A fine toothed back saw does an excellent job. It can also be turned on a lath, but keep the cutting speed low. It can be worked on a milling machine. Again, keep the cutting speed low. It can be worked with a carving bit in a "moto-tool." It can also be drilled like wood and can be tapped as long as you use fairly coarse threads. The size $1/4 \times 20$ works well.

The ugly black markings can be removed with lacquer thinner, but that may leave a slightly dulled surface. You can paint it with any good enamel, but the color is subject to scratching.

The inside of PVC is shiny and reflects light. If used as a telescope tube it must be painted or lined with "flocking cloth." If a section is too long to be easily spray painted from each end you can tape the spray can to a dowel and rig a way of activating the trigger with a string or another dowel. You aren't trying to get a neat paint job. No one, including you, can see it inside the tube. You are just trying to get flat black paint everywhere.

PVC does not slide very well against itself. Tight telescoping tubes tend to grab and stick. The threads and slip joints are, like standard plumbing pipe, tapered. That makes them loose at the beginning and grabby at the end. Unlike cast iron pipe threads you can't lap them together to get a better fit. You can use milk carton plastic to provide a bearing surface if you have enough clearance between the surfaces. Or use thick grease on PVC threads etc. to get them to move reasonably smoothly. The only complication to greasing is the fact that the grease tends to migrate to other locations such as lens surfaces.

PVC can be shaped with a coarse file and worked with a disk sander, but it tends to fill the grit of the abrasive paper. A disk sander is excellent for getting square ends and flat surfaces. Keep the cutting speeds moderate. You don't want to melt the surface.

PVC can be bent into useful shapes after heating it. You can immerse it in boiling water on the stove, then force it into the desired shape with a pair of pliers, or into or around a mold (wood). You can also heat it in an oven. Be careful that you don't start a fire or melt it all over the inside of the oven. You can get small sheets to work with by sawing through the side of a piece cut off from a pipe. A piece of 6-inch pipe will give you a piece more than 18 inches on one side and as long as the piece of pipe. See www.geocities.com/napsp2000/focuser.html for a PVC-based Crayford type focuser.

Obviously you need to be careful. Don't get burned.

Cut 90-degree joints at a 45-degree angle. Then twist to fit the cut edges to make a 90 degree bend. Carefully finish the edges to get good contact and a true 90-degree bend. A disk sander is ideal for the edge finishing. Apply PVC cement to each edge and hold the edges together. The joint will set quite quickly. Figure 5.1 (Chap. 5) shows a right angle joint used for a star diagonal. The focuser is mounted on a wooden plug. The focuser was purchased online from a large manufacturer of telescopes.

PVC Eyepiece Canisters

Eyepieces get knocked off of whatever they are placed on. They land in the grass and, likely, get stepped on. Sticking an unused eyepiece in a pocket is not a good solution because pocket lint does not improve eyepiece performance. It is easy and cheap to make some nice little canisters. You still may knock them into the grass and step on them, but you can't break the eyepiece.

For N Canisters

You need enough PVC to make N pieces at least as long as your longest eyepiece. A 1 1/2 inch inside diameter works well for standard 1 1/4 diameter eyepieces. If you use larger eyepieces obviously adjust. You need N PVC caps to fit. The caps will be the removable tops – enough scrap wood to turn a 1 7/8 diameter dowel. (Glue up some 3/4 stock if you like.) If you don't have a lathe some scrap 1/4 or 3/8 inch plywood will work. Just make two disks for each canister, one to match the inside diameter and one to match the outside diameter. Glue them together, and you have plugs for the bottoms for your canisters.

If you have, or can gain access to, a wood lathe turn a dowel to the same diameter as the outside diameter and a step down to the inside diameter. Saw the turned piece to make a bunch of plugs for the bottoms of your canisters. Construction cement will hold them in place nicely. PVC cement doesn't adhere to the wood plugs very well.

Other Uses for PVC

Aligning Parts

Don't restrict your thinking about PVC to situations in which it will be the primary material or serve as a tube. It can be very useful as a way of aligning parts. A 4-inch refractor was designed to allow easy twisting of the star diagonal for user comfort and/or its replacement with a camera.

It is obvious that when a change is made alignment needs to be automatic. The solution was to build two carefully aligned rings into the main tube and to equip the matching end of the diagonal, camera, or any other thing to be attached at the back end of the telescope, with a piece of PVC to slip into the rings. To get around the fact that PVC slip joints are tapered the rings were sawed from the outboard ends of a PVC coupling. Figure 13.1 shows the coupling and the backsaw. The coupling happens to be made for a 12 1/2 degree joint, but this is of no significance, since the entire center part will be trashed. Only the outboard 1/4 or a little less is actually used. It may be necessary to relieve the inner edges a little with mototool. Don't attempt to use any more, because even a very slight tilt will make them grab if they have much width. They serve to keep the whole diagonal square because they are separated by about 6 inches.

Fig. 13.1 PVC Coupling to be sawed into retaining rings

The objective lens cell for the same telescope was made from a straight coupling for 4-inch PVC. The flange in the middle of the coupling needed to be cleaned up to allow the lens to fit flat. A thin cutoff from a 4-inch PVC tube served as a retaining ring to hold the lens in place. Two machine screws threaded into the cell put a squeeze on the ring, distorting it enough to stay in place. The inside had to be slightly relieved with the sanding attachment on a mototool to make it slip onto a piece of 4-inch PVC that was used to guarantee alignment.

Telescope Mounts

PVC is also very effective as a pier for a telescope mounting. Just make a hole, stick the pipe into the hole, and fill both the hole and the pipe with concrete. Use a 6-inch pipe as a minimum. See the section on a flexible, universal mounting. If the hole is reasonably deep, say 24 inches, you won't need to brace it to keep it square while the concrete is setting. Just push it until the square reads correctly, wiggle it a little to make sure that the concrete is well settled against the base of the pipe, and just let go. It should just stand there. It is probably a good idea to have a couple of stakes and some string handy in case your concrete mix is soupy and the whole thing starts to slump.

Lens Cells

Some of the lenses that you may get from web sources may have no cell or may have a cell that just doesn't accommodate itself to mounting in a tube. Some perfectly usable lenses come as part of some strange, non-symmetrical or oversized contraption. You can usually get it out of its mount, perhaps by sawing, etc. The result is that you have a fine lens with no cell.

Retaining Rings

If you can fit the lens into some standard size PVC tube you can make retaining rings from the same size tube. Cut a convenient length and then cut a slot down one side. The slot will allow you to compress the ring and get it inside the uncut piece of the same stuff. You need to take out a pretty good-sized slice to get enough space. Many times the wall thickness is good. It the slot is too wide and doesn't quite close no harm is done.

Allow the piece that you are using as a ring to be reasonably long. Two inches is not excessively long for the inner ring, because it will be the one against which the lens rests. That makes it much easier to get it squarely placed in the outer tube. See Fig. 13.2. It shows a long ring on which the lens will rest, a short ring that will only hold the lens in place, and the rather long tube into which the lens and rings will be placed as a completed cell. A long cell is often easier to mount squarely in the telescope tube.

You may find that it is difficult to apply enough pressure to the split rings to get them to fit into the cell. If a vice does not apply an even enough pressure you can use a hose clamp – not a very different process than getting new piston rings into a cylinder. Once you get the inner tubes inside the outer tube you need a way of keeping them in place. The one that supports the lens needs to be placed permanently, and the one that serves as a retaining ring must be removable. For the permanent ring PVC cement is inappropriate because it sets too quickly. A good solution is to

Fig. 13.2 Parts for an Objective cell

drill through both the outer and inner pieces, tap the hole, and use a short machine screw. For the retaining ring just drill and tap the outer tube and use a machine screw pressing against the inner ring.

Floor Underlay

The material sold as an underlay for floors is very useful and cheap. It is slightly less than 1/4 of an inch thick and normally has only three layers of wood rather than the four or more in regular plywood. The most common surface wood is "Luann," or "Philippine mahogany." This is a rather soft, open grain wood that can be made into uniform sheets easily. Hence its use in the underlay material. It needs a primer – filler to paint well and looks attractive, but plain if varnished. It is easy to work with, is made with a water-proof glue, and is ideal for box-type telescope tubes. Since it is rather thin and not very strong it needs a significant amount of internal bracing.

Glues and Cements

Tremendous progress has been made in this area in the last few years regarding stuff you can use to stick other stuff to. What follows is just a sampling. Go to your local hardware supplier and spend some time in the adhesives aisle.

Wood Glues	There are fairly quick setting and slow setting varieties. The slow-setting varieties are very useful for projects with many joints that need to be pushed around after they all have been put together, wood telescope tubes, for example. There are many manufacturers
Construction Glues	Made in a dazzling variety of types. Useful for wood construction, to attach laminates to wood, etc. This stuff comes in tubes to be used with glue guns. It is pretty thick and gooey, so it can be used to attach things to surfaces not completely smooth. Some varieties have a lot of initial tackiness, so things will stay together before the glue sets. There are many manufacturers
Epoxy	Also, a dazzling variety of types. Some set very quickly, some more slowly. These glues require mixing just prior to use
J-B Weld	This also requires mixing just prior to use. Sets quite slowly but extremely strong. Works for such telescope applications as metal to metal attachment of camera lenses to eyepiece tubes. Don't use this stuff in the wrong place. It is hard to get it loose
Pitch	Pitch is definitely not a modern adhesive. It has been around for thousands of years and some anthropologists feel it may pre-date all other adhesives. In optics it is commonly used to make polishing tools. Other neat applications are where you want to have a strong but temporary connection. Under pressure it is very strong, but it yields very easily to shock. A tap with a hammer can separate a joint that just sheer force won't touch. You have to melt the stuff to use it and it smells like a pine tree – which is, after all, where it comes from

Appendix I

Web Suppliers of Telescope Building Parts and Materials

No endorsement or recommendation is provided or implied. The organizations listed are here because they were found in other such lists or advertisements or were personally known to someone. Those for whom e-mail addresses were available were notified that they would be listed. Those for whom there were no e-mail addresses or whose e-mail was not deliverable were sent regular letters with the same information. If there was no response to notification, the listing was removed.

Quite a few listed organizations submitted additions or changes to their list of products. References to items specific to astrophotography and to Dobsonians were, for the most part, avoided. Likewise, references to subjects other than building telescopes, such as star charts, were avoided.

If you are searching for some particular item or for something that will serve for a particular item you are advised to look under a variety of related categories. There is very little standardization of nomenclature in telescope design. For example, "used" usually means "buys and sells used parts and complete scopes." Rings and mounting rings probably are overlapping categories. Diagonals, mirrors, and secondary mirrors probably all mean the same thing.

Some of the listed organizations produce or distribute items useful to us telescope makers but with a much wider/different selection of clients. In dealing with them it would be useful if you are particularly clear as to the intended use. Some produce items for other industries and are set up to supply materials in large quantities. Be very courteous to them if you expect them to accommodate tiny orders.

This appendix contains three types of listings:

- an alphabetical list of suppliers
- a list of all the product designations
- a cross-referenced listing of suppliers by product

R.L. Clark, *Amateur Telescope Making in the Internet Age*, Patrick Moore's
Practical Astronomy Series, DOI 10.1007/978-1-4419-6415-1,
© Springer Science+Business Media, LLC 2011

Alphabetical Listing of Suppliers

The format of these listings is:

   Organization or company name (with ID number for cross reference)

   Mailing address (if one is available)

   Website address contact   e-mail phone number (if available)

   Products are separated by a (/)

1800destiny.com (1)
PO Box 5191 "Pleasanton CA 94566"
www.destiny.com   hans@destiny.com   800-337-8469
coating services/focusers/diagonals/spiders/mirrors/

ADM Accessories (2)
www.admaccessories.com   sales@admaccessories
mounting rings/mounts/dovetails/

Adorama (3)
42 West 18th Street "New York, NY 10011"
www.adorama.com   info@adorama.com   212-741-0052
eyepieces/diagonals/

Aeroquest (4)
HCO2 Box 7334E "Palmer, AK 99645"
www.aeroquest-engineering.com   aeroqst@mtaonline.net   907-745-4091
machining services/drives/

Agena AstroProducts (5)
10805 Artesia Blvd # 109 "Cerritos, CA 90703"
www.agenaastro.com   info@agenaastro.com
562-972-3738 9 AM–8 PM Pacific Time
eyepieces/focusers/adapters/filters/

Alpha Optics (7)
alphaprecisionoptics@yahoo.com   Terry@alphaprecisionoptics@yahoo.com
diagonals/

Alpine Astronomical (8)
PO Box 1154 "Eagle, ID 83616"
www.alpineastro.com   sales@alpineastro.com   208-939-2141
adapters/rings/finders/mounts/filters/diagonals/binoviewers/barlows/

Aluminum Coating (9)
807 Rutherdale "San Carlos, CA 94070"
http//covad.net/-alcoat alcoat@covad.net
coating services/

American Science & Surplus (10)
PO Box 1030 "Skokie, IL 60076"
www.sciplus.com   info@sciplus.com   888-SCIPLUS
beamsplitters/lenses/mirrors/filters/prisms/projection lenses/

Anacortes (11)
9973 Padilla Heights Rd "Anacortes, WA 98221"
www.buytelescopes.com   info@buytelescopes.com   800-850-2001
finders/eyepieces/filters/barlows/eyepieces/mounts/

Apogee Components (12)
3355 Filmore Ridge Heights "Colorado Springs, CO 80907"
www.apogeerockets.com   info@apogeerockets.com   719-535-9335
tubes/

Astro Engineering (14)
www.astro-engineering.com   info@astro-engineering.com   866-842-3925
eyepieces/

Astro Optics (14)
S/3 Kapole Soc. 13th cross Rd, Juhu Mumbai – 400049, Maharashta, India
www.astro-optical.com   info@astro-optical.com   91 22 26118942
mirrors/beamsplitters/prisms/filters/flats/coating services/

Astro Parts Outlet (15)
P.O. Box 1873 "Frazier Park, CA 93225"
www.astronomy-mall.com   or   www.cloudynights.com   or   www.astromart.com
newyearskid@webtv.net   or   newyearsskid@att.net   661-245-0805   or
818-501-6920
mounts/rings/focusers/finders/mirrors/lenses/diagonals/spiders/prisms/
barlows/binoviewers/adapters/filters/

Astro Sky (16)
3705 "Swanee St Lake Charles LA 70607"
astrosky@homestead.com   astrosky@homestead.com   337-564-5482
finders/flocking/dew shields/filters/shroud material/cells/mirrors/diagonals/

Astro Systems (17)
124 N. 2nd St "LaSalle, CO 80645"
www.astrosystems.com   info@astrosystems.com   970-284-9471
spiders/mirror cells/dew shields/diagonals/focusers/mirrors/spiders/

Astromart Auctions (18)
www.astromart.com   info@astromart.com/auctions
used/

Astromart Classafied (19)
www.astromart.com   info@astromart.com/classifieds
used/

Astronomics (20)
680 24th Ave SW "Norman, OK 73069"
www.astronomics.com   questions@astronomics.com   800-422-7876
diagonals/filters/eyepieces/

Astronomik (21)
Eiffestrase 426 D-20537 Hamburg GR
www.astronomic.com   astro@astro-shop.com   +49(0)40/511 43 48
filters/

Astro-Physics (22)
11250 Forest Hills Rd "Machesny Park, IL 61115"
www.astro-physics.com   info@astro-physics.com   815-282-1513
eyepieces/barlows/diagonals/binoviewers/filters/focusers/

Astrotelescopes (23)
www.astrotelescopes.com   info@astrotelescopes.com
Astrozap (24)

PO Box 502 "Lakewood, OH 44107"
www.astrozap.com sales@astrozap.com 212-334-1622
used/dew shields/filters/heaters/artificial stars/

Aurora Astro Products (25)
11419 19th Ave SE #A102 "Everett, WA 98208"
www.auroraastro.com   info@auroraastro.com
eyepieces/piggy-back mounts/diagonals/mounts/adapters/filters/setting
circles/focusers/finders/

B & H (26)
420 9th Ave "New York, NY 10001"
www.bhphotovidio   info@bhphotovidio   800-606-6969
eyepieces/barlows/diagonals/mirrors/finders/rings/

BigEye Optics (27)
www.digital-flight.com/thebigeye/optics.html   acgna@comcast.net
re-figuring services/mirrors/

Bob's Knobs (28)
6978 Kempton Rd "Centerville, IN 47330"
www.bobsknobs.com   info@bobsknobs.com
knobs/

Boulter Plywood (29)
24 Broadway "Somerville, MA 02145"
boulterplywood.com   info@boulterplywood.com   888-9 lumber
special wood/

Brunner Enterprises (30)
455 Center Rd "West Seneca, NY 14224"
www.brunnerent.com   info@brunnerent.com   877-299-2622
tubes/

Burgess Optical (31)
PO Box 32515 "Knoxville, TN 37930"
www.burgessoptical.com   info@burgessoptical.com   865-769-8777
eyepieces/binoviewers/mounts/finders/diagonals/barlows/filters/

Bushnell (32)
9200 Cody "Overland Park, KS 66214"
www.bushnell.com/astronomy   info@bushnell.com/astronomy   800-423-3537
eyepieces/adapters/erecting prisms/barlows/

Byers Co. (33)
290001 W. Hwy 58 "Barstow, CA 92311"
erbyers@gte.net   erbyers@gte.net   760-256-2733

Celestron (34)
2835 Columbia St "Torrance, CA 90503"
www.celestron.com/c3/homr.php   info@celestron.com/c3/homr.php
800-447-1001
eyepieces/barlows/t-Rings/filters/

Chicago Tube (35)
One Chicago Tube Dr "Romeoville, IL 60446"
www.chicagotube.com   info@chicagotube.com   815-588-3958
tubes/

Clausing (36)
8038 N. Monticello Ave "Skokie, IL 60076"
www.clausing.com   info@clausing.com   847-676-0330
coating services/

Clement (37)
PO Box 16 "Running Springs, CA 92382"
www.clementfocuser.com   info@clementfocuser.com   909-867-4519
focusers/

Commercial Sales Inc. (38)
6411 Pacific Highway E "Tacoma, WA 98424"
www.commercialsalesinc.com   info@commercialsalesinc.com   253-922-6670
used portholes/

Company Seven (39)
Box 2587 "Montpelier, MD 20709"
www.company7.com   info@company7.com   301-953-2000

Compact Precision Telescopes (40)
408-947-1968
mirror cells/

Coronado (41)
"1674 Research Lp, # 436" "Tucson, AZ 85710"
www.coronadofilters.com   info@coronadofilters.com   520-740-1561
filters/

Craftflocking (42)
www.craftflocking.com   info@craftflocking.com
flocking paper/flocking fibers/adhesives/flocking guns/

DAR Astro Machining (43)
www.onellphoto.on.ca   info@onellphoto.on.ca   519-679-8840
adapters/focusers/cells/

David Lukehurst (44)
87 Victoria Rd, Sherwood "Nottingham NG5 2NL, UK"
www.dobsonians.co.uk   info@dobsonians.co.uk   00 44 (0)115 9602266
focusers/mirrors/eyepieces/barlows/diagonals/finders/

DayStar Filters (45)
149 Northwest OO "Warrensburg, MO 64093"
www.daystarfilyers.com   service@datstarfilters.com   866-680-6563
filters/

Del Mar Ventures (46)
12595 Ruette Alliante #148 "San Diego, CA 92130"
www.sciner.com/opticsland   info@sciner.com/opticsland   509-752-0123
blanks/lenses/prisms/beamsplitters/

Denkmeier Optical (47)
12623 Sunset Ave #4 "Ocean City, MD 21842"
deepskybinoviewer@mchsi.com   866-340-4578
eyepieces/binoviewers/barlows/

Denton Vacum (48)
1259 North Church St "Morrestown, NJ 08057"
www.dentonvacuum.com   info@dentonvacuum.com   856-439-9100
coating services/

DiscMounts (49)
www.discmounts.com   discmounts.com   954-475-8574
mountings/tripods/

Dream Telescopes (50)
www.dreamscopes.com   info@dreamscopes.com   610-365-2833
tubes/cells/

Edmund Scientific (51)
101 East Gloucester Pike "Barrington, NJ 08007"
www.edmundscientifics.com   info@edmundoptics.com   1-800-363-1992
coating services/rings/lens cells/filters/lenses/prisms/flocking paper/barlows/

EfstonScience (52)
3350 Dufferin St "Toronto, ON Canada M6A 3A4"
www.escience.ca   info@escience.ca   416-787-4581
eyepieces/barlows/adapters/mirrors/diagonals/dew shields/

EMF (53)
239 Cherry St "Ithica, NY 14850"
www.emf-corp.com   info@emf-corp.com   607-272-3320
coating services/

Epsilon Telescopes (54)
"Penmanor Obs, 56 Penmanor" "Finstall, Bromsgrove, Worcs B60 3Bz"
www.epsilon-telescopes.co.uk   info@epsilon-telescopes.co.uk   01527 880431
adapters/rings/machining services/

E-Scopes (55)
13860-38 Wellington Trace #111 "Wellington, FL 33414"
www.e-scopes.cc   murni@gate.net   561-795-2201
diagonals/mirrors/spiders/focusers/barlows/filters/finders/

FAR Labs (57)
www.dynapod.com   info@dynapod.com   800-336-9054
mounts/

Finger Lakes Instrumentation (58)
www.flcamera.com   info@flcamera.com

Galaxy Optics (59)
PO Box 2045 "Buena Vista, CO 81211"
www.galaxtoptics.com   info@galaxtoptics.com   719-395-824
mirrors/diagonals/coating services/

Galvoptics (60)
Harvey Road, Burnt Mills Ind Estate "Basildon, Essex UK SS13 1ES"
www.galvoptics.fsnet.co.uk   inf@galvoptics.fsnet.co.uk   +44(0)1268 728077
mirrors/mirror kits/abrasives/flats/diagonals/filters/coating services/

Glass Fab (61)
PO Box 31880 "Rochester, NY 14603"
www.glassfab.com   info@glassfab.com   585-262-4000
blanks/

Ganymede Optics (62)
PO Box 141 "Barrington, NJ 08007"
www.ganymeadoptics.com   info@ganymeadoptics.com   856-939-8951
eyepieces/barlows/

GotGrit (63)
2612 Maylin Drive "Trinity, FL 34655"
www.gotgrit.com   info@gotgrit.com
abrasives/blanks/pitch/grinding kits/

Great Red Spot Astronomy Products (64)
15805 Colbert St "Romulus, MI 48174"
www.greatredspot.com   info@greatredspot.com   734-532-2818
eyepieces/mounts/filters/dew shields/

Hancock Fabrics (65)
218 N. Fredrick Ave "Gaithersburg, MD 20877"
www.hancockfabrics.com   info@hancockfabrics.com   301-977-2733
flocking fabric/

Hands on Optics (66)
26437 Ridge Road "Damascus, MD 20872"
www.handsonoptics.com   astroguy@handsonoptics.com   301-482-0000
eyepieces/diagonals/used/filters/barlows/adapters/mirrors/lenses/
mounts/spiders/mount parts/

Harrison Telescopes Ltd (67)
www.harrisontelescopes.co.uk   info@harrisontelescopes.co.uk
0844 8801377-01932 703605
eyepieces/barlows/diagonals/focusers/mounts/

Hastings Irrigation Pipe (68)
PO Box 728 "Hastings, NE 68902"
www.hipco-ne.com   info@hipco-ne.com   402-463-6633
tubes/

Helix (69)
PO Box 490 "Gibsonia, PA 15044"
www.helix-mfg.com/hercules.htmjthagan@helix-mfg.com   info@helix-mfg.com
724-316-0306
eyepieces/barlows/mounts/

Hextek (70)
1665 E. 18th St #208 "Tuscon, AZ 85718"
www.hextek.com   info@hextek.com   520-623-7647
blanks/

High Point Scientific (71)
442 Route 206 "Montague, NJ 07827"
www.highpointscientific.com   info@highpointscientific.com   800-266-9590
eyepieces/mounts/barlows/diagonals/filters/used/

Hobbylink (72)
www.hobbylink.com/rochets/rockets.htm   hobbylink.com
tubes/

Hotech (73)
9320 Santa Anita Ave #100 "Rancho Cucamonga, CA 91730"
www.hotechusa.com   info@hotechusa.com   909-987-8828
adapters/laser collimators/green lasers/

Hubble Optics (74)
www.hubble-optics.com   info@hubble-optics.com
mirrors/artificial stars/diagonals/

Intermountain Optics (75)
ioptics@xmission.com   ioptics@xmission.com   801-651-0531
coating services/mirrors/blanks/diagonals/

Island eyepiece and Telescope (76)
PO Box 133 "Mill Bay, BC VOR 2PO, Canada"
www.islandeyepiece.com   info@islandeyepiece.com   250-743-6633
eyepieces/mirrors/rings/dew shields/binoviewers/diagonals/filters/finders/mounts/

ISP Optics (77)
1 Bridge St "Irvington, NY 10533"
www.ispoptics.com   info@ispoptics.com   914-591-3070
beamsplitters/lenses/mirrors/prisms/

ITE (78)
16222 133rd Drive "Jupiter, FL 33478"

www.iteastronomy.com   mike@iteastronomy.com   800-699-0906
used/adapters/barlows/diagonals/eyepieces/focusers/mounts/

JMI Telescopes (79)
8550 West 14th Ave "Lakewood, CO 80215"
www.jimsmobile.com/   info@jmitelescopes.com   303-233-5353
focusers/dollies/cary cases/focus motor controls/tracking motors/

JoeAstro (80)
mysite.verizon.net   joeastro@aol.com   760-373-4290
cells/finders/rings/spiders/

Kahn Scope Centre (81)
3243 Dufferin St "Toronto, ONT Canada M6A 2T2"
www.kahnscope.com   info@kahnscope.com   800-580-7160
eyepieces/barlows/filters/adapters/finders/mounts/dew shields/diagonals/focusers/

Kendrick Astro Instruments (82)
36 Cawthra Ave "Toronto, ONT M6N 8B3"
www.kendrickastro.com   info@kendrickastro.com   416-762-7946
dew shields/observer tents/used/

Kennedy Optics (83)
PO Box 994 "Yucca Valley, CA 92286"
www.kennedy-optics.com   steve@kennedy-optics.com   760-369-2206
large mirrors/

Kens Rings (84)
www.kendauzat.com   info@kendauzat.com
rings/mounts/

Kine Optics (85)
451 Elk Loop "Sequim, WA 98382"
www.kineoptics.com   info@kineoptics.com   360-406-4019
focusers/

L & L Optical Services (86)
23352 Madero Rd #A "Mission Viejo, CA 92691"
www.llopt.com   info@llopt.com   949-470-9672
coating services/

Le Suer (87)
3220 Lorna Rd "Birhmingham, AL 35216"
www.astropier.com   info@astropier.com   205-822-0720
piers/

Lightholder Optics (88)
PO Box 14271 S "Lake Tahoe, CA 961521"
www.lightholderoptics.com   info@lightholderoptics.com   530-577-4328
mirrors/

Lockwood Custom Optics (89)
909 W White St "Champaign, IL 61821"
www.loptics.com   info@loptics.com
mirrors/

Lookinglass (90)
www.embeddedrf.com   info@embeddedrf.com
blanks/gears/

Lumicon (91)
750 Easy St "Simi Valley, CA 93065"
www.lumicon.com   info@lumicon.com   805-520-0047
filters/

Magee Optics (92)
27 Southfield Rd "Concord, MA 01742"
rmagee@bitnet.net   rmagee@bitnet.net

Maier Photonics (93)
maierphototonics.com   maierphototonics.com   802-362-1042
filters/beamsplitters/

Majestic Optical Coatings (94)
152 Willow Way "Clark, NJ 07066"
www.majestic-coatings.com   info@majestic-coatings.com   732-388-5604
coating services/

Markless (95)
www.marklessastronomics.com   info@marklessastronomics.com
dob accessories/

Mathis (96)
www.mathis-instruments.com   info@mathis-instruments.com   925-838-0444
mountings/

Meade (97)
27 Hubble "Irvine, CA 92618"
www.meade.com   info@meade.com   800-626-3233
filters/eyepieces/barlows/

Melles Griot (98)
200 Dorado Place SE "Albuqueque, NM 87123"
optics.mellesgriot.com   info@optics.mellesgriot.com   505-296-9541
mirrors/shutters/lenses/prisms/flats/beamsplitters/filters/

Micro Abrasives (99)
PO Box 669 "Westfield, MA 01086"
www.microgrit.com   info@microgrit.com   800-426-6046
abrasives/

Micro Sphere (100)
11 impasse des carieres 58180 "MARZY, France"
franck.griere@orange.fr   franck.griere@orange.fr   03 86 61 37 33
mirror kits/mirrors/

Mirror Making Material (101)
Kromptestr 6 D-81543 "Munich, Germany"
www.stathis-firstlight.de/english   info@stathis-firstlight.de/english   8.82E-09
blanks/abrasives/

MoonLite (102)
114 Ardmoor Ave "Danville, PA 17821"
www.focuser.com   info@focuser.com   570-275-7935
focusers/

New Rise Optics (103)
1890 Marich Way "Mountain View, CA 94040"
www.newrise-llc.com   info@newrise-llc.com   650-948-0232
lenses/prisms/coating services/beamsplitters/

Newport Glass (104)
10564 Fern Ave "Stanton, CA 90680"
www.newportglass.com   info@newportglass.com   714-484-8100
mirror kits/lens blanks/mirrors/abrasives/lens kits/blanks/

Newsom Precision Telescope Mirrors (105)
PO Box 207 "Driggs, ID 83422"
Newsomparabolics@gmail.com   Newsomparabolics@gmail.com
208-354-2734
mirrors/pyrex/quartz/

Nichol Optical (106)
Northcote Hill Frmhs (off) Rd, "Norton, Stockton-on-Tees Ts201LB UK"
www.nicholoptical.co.uk   info@nicholoptical.co.uk   01642 554699
mirrors/

Norman Fullum (107)
77 Sanderson, Hudson "Quebec, Canada J0P 1H0"
www.normanfullumtelescope.com   info@normanfullumtelescope.com
514-967-1909

mirrors/re-figuring services/
Nova Optical Systems (108)
14121 S Shaggy Mountain Cir "Herriman, UT 84096"
www.nova-optical.com   info@nova-optical.com   801-446-1802
coating services/flats/

Oldham Optical UK (109)
10 Royal Crescent Lane "Scarborough North Yorkshire, UK YO11 2RL"
oldham-optical.co.uk   info@oldham-optical.co.uk   044(0)1723506050
blanks/diagonals/secondaries/flats/

Optcorp (110)
www.optcorp.com   info@optcorp.com   800-483-6287
eyepieces/barlows/filters/extenders/finders/

Optic Wave Labs (111)
11358 Amalgam Way, Suite A2 "Gold River, CA 95670"
www.opticwavelabs.com   cary@Opticwavelabs.com   916-671-5962
parabolic mirrors/spherical mirrors/flats/coating services/blanks/refiguring services/

Optical Mechanics (112)
PO Box 2313 "Iowa City, IA 52244"
www.opticalmechanics.com   info@opticalmechanics.com   319-351-3960
mirrors/coating services/

Optical Vision Ltd (113)
#3 Woolpit Bus Park Woolpit, Bury St "Edmunds, Suffolk IP30 9UP, England"
www.opticalvision.com   info@opticalvision.com
eyepieces/barlows/filters/diagonals/binoviewers/finders/focusers/adapters/focusers/

Optics Planet (114)
3150 Commercial Ave "Northbrook, IL 60062"
www.opticsplanet.com   sales@opticsplanet.com   800-504-5897
eyepieces/filters/rings/adapters/mounts/

Orion (115)
89 Hanger Way "Watsonville, CA 95076"
www.telescope.com/control/main/   sales@telescope.com   800-676-1343
barlows/diagonals/filters/eyepieces/focusers/

Owl (116)
5950 Keystone Dr "Bath, PA 18014"
www.Owlastronomy.com   info@Owlastronomy.com   888-Owl-ware
eyepieces/

Parabolized Telescope Mirrors (117)
mrmaker07.tripod.com   planetsurfur@yahoo.com   540-476-1221
mirrors/

Parks (118)
750 E Easy St "Simi Valley, CA 93065"
www.parksoptcal.com   info@parksoptcal.com   805-522-6722
eyepieces/tubes/diagonals/mirrors/blanks/spiders/focusers/rings/setting
circles/filters/

Pegasus Optics (119)
16426 Martins Ferry "San Antonio, TX 78247"
Pegasusoptics.com   info@Pegasusoptics.com   830-538-9499
mirrors/

Peterson Engineering (120)
405 New Meadow Rd "Barrington, NJ 02806"
www.petersonengineering.com   info@petersonengineering.com   401-245-6679
adapters/gears/clutches/

Pelican Wire (121)
3650 Shaw Blvd "Naples, FL 34117"
www.pelicanwire.com   info@pelicanwire.com   239-597-8555
heat rope/

Pines Optical (122)
5429 Overlook Rd "Milford, OH 45150"
www.pinesoptical.com   pinesop@aol.com   513-831-7045
mirrors/

Precise Parts (123)
www.preciseparts.com   info@preciseparts.com   305-253-5707
machining services/

Precision Applied Optics (124)
355-B Crowther Ave "Placentia, CA 92870"
www.prescisionapplied.com   info@prescisionapplied.com   714-579-7812
mirrors/lenses/beamsplitters/prisms/coating services/

ProtoStar (125)
PO Box 448 "Worthington, OH 43085"
www.fpi-protostar.com   info@fpi-protostar.com   866-227-6240
spiders/diagonals/flocking/tubes/

Public Missiles (126)
6030 Paver Lane "Jeddo, MI 48032"
publicmissiles.com   info@publicmissiles.com   810-327-1710
tubes/

Pulsar-Optical (127)
www.pulsar-optical.co.uk   info@pulsar-optical.co.uk
eyepieces/

R.F. Royce Precision Optical Components (128)
30 Holly Mar Hill Rd "Northford, CT 06472"
www.rfroyce.com   info@rfroyce.com   203-484-7705
mirrors/

Reginato (129)
Italy
info@reginato.it
blanks/glass working services/mirrors/optical testing services/

Research Service (130)
30-B Revere Beach Pkwy "Revere, MA 02151"
glasscoater.com   info@glasscoater.com   781-284-0303
coating services/

Rigel Systems (131)
26850 Basswood Ave "Rancho Palos Verdes, CA 90275"
www.rigelsys.com   info@rigelsys.com   310-375-4149
finders/adapters/finders/

Robert Clark Optical Glass (132)
49 Fitzhugh Ave "Westminster, MD 21157"
www.rlclark21157@gmail.com   rlclark21157@gmail.com   410-751-9710
lens blanks/raw optical glass/

Rolyn Optics (133)
706 Arrowgrand Circle "Covina, CA 91722"
www.rolyn.com   info@rolyn.com   888-626-1379
lenses/prisms/flats/beamsplitters/

ScopeStuff (134)
PO Box 3754 "Cedar Park, TX 78630"
www.scopestuff.com   support@scopestuff.com   512-259-9778
mounting rings/focusers/eyepieces/spiders/diagonals/dob
parts/mounts/t-adapter/finders/filters/heaters/shutters/

ScopeTronix (135)
www.scopetronix.com   info@scopetronix.com
mirrors/diagonals/adapters/focusers/

SCS Astro (136)
1 Tone Hill "Wellington Somerset, TA21 0AU, UK"
www.scsastro.co.uk   info@scsastro.co.uk   +44(0)1823 665510
eyepieces/mounts/focusers/filters/mounts/barlows/extenders/diagonals/
adapters/used/

Seymour Solar (137)
61637 Vega St "Bend, OR 97701"
www.seymour.com   info@seymourcolar.com   541-306-6178
filters/

Shapiro Supply (138)
5617 Natural Bridge "St Louis, MO 63120"
www.shapirosupply.com   info@shapirosupply.com   800-833-1259
tubes/

SkySpot (140)
1263 East Beverly Way "Bountiful, UT 84010"
www.sky-spot.com   info@sky-spot.com   800-363-9540
finders/

Small Parts Inc. (142)
15901 SW 29 Street #201 "Miramar, FL 33027"
www.smallparts.com   info@smallparts.com   800-2204242
tubes/aluminum strips-bars/brass strips-bars/taps/dies/tools/

Spectrum Coatings (143)
1165 Ring St "Deltona, FL 32725"
www.spectrum-coatings.com   info@spectrum-coatings.com   386-561-9779
coating services/

Starlight Instruments (144)
2380 E Cardinal Drive "Columbia City, IN 46725"
www.starlightinstruments.com   info@starlightinstruments.com   260-244-0020
focusers/adapters/

Starmaster Telescopes (145)
2160 Birch Rd "Arcadia, KS 66711"
www.starmastertelescopes.com   starmaster@ckt.net   620-638-4743
focusers/mirror heaters/

Stellavue (146)
11820 Kemper Rd "Auburn, CA 95603"
www.stellarvue.com   vic@stellavue.com   530-823-7796
eyepieces/diagonals/finders/binoviewers/mounting
hardware/guidescope rings/barlows/finders/

Sterling Resale Optics (147)
www.sro-optics.com   info@sro-optics.com
lenses/mirrors/filters/beamsplitters/

Stevens Optical (148)
7335 SE 19th Ave "Portland, OR 97202"
www.stevensoptical   info@stevensoptical   503-203-8603
mirrors/re-figuring services/

Steller Technologies (149)
www.stellar-international.com   info@stellar-international.com   800-232-9416
focusers/

Surplus Shed (150)
1050 Maidencreek Rd "Fleetwood, PA 19522"
www.surplusshack.com   sales@surplusshed.com   1-877-7surplus(78-7758)
achromats/lenses/filters/flats/blanks/prisms/diagonals/objective lenses/

T&A (151)
Box 292 "Ohalloran Hill, SA 5158 Australia"
www.telescopes-astronomy.com.au   info@telescopes-astronomy.com.au
61'+8 8381 3188
eyepieces/mirrors/barlows/diagonals/lenses/focusers/adapters/mounts/

Tech2000 (152)
69 Ridge St South "Monroeville, OH 44847"
www.tech2000astronomy.com   info@tech2000astronomy.com   419-465-2997
focusers/dew shields/mounts/

Teeters Telescopes (153)
"Lake Hiawatha, NJ 07034"
www.teeterstelescopes.com   rob@teeterstelescopes.com   732-991-1248
mirrors/diagonals/cells/spiders/focusers/

Televue (154)
www.televue.com   info@televue.com
eyepieces/extenders/filters/mounts/

The Science Company (155)
95 Lincoln St "Denver, CO 80203"
secure.sciencecompany.com   info@secure.sciencecompany.com   800-372-6726
eyepieces/barlows/filters/finders/prisms/

Thousand Oaks Optical (156)
Box 6354 "Kingman, AZ 86401"
www.thousandoaksoptical.com   info@thousandoaksoptical.com   928-692-8903
filters/dew control systems/

University Optics (157)
PO Box 1205 "Ann Arbor, MI 48106"
universityoptics.com   uoptics@aol.com   734-665-3575
eyepieces/

Van Slyke (158)
12815 Porcupine Ln "Colorado Springs, CO 80908"
www.observatory.org   info@observatory.org   719-495-3828
focusers/

Velocity Air (159)
stores.shop.ebay.com/Velocity-Air__W0QQ_armsZ1
info@stores.shop.ebay.com/Velocity-Air__W0QQ_armsZ1
tubes/

Vernon (160)
5 Ithica Rd "Candor, NY 13743"
www.vernonscope.com   info@vernonscope.com   607-659-7000
diagonals/

Vixen Optics (161)
1010 Calle Cordillera #106 "San Clemente, CA 92673"
www.vixonoptics.com   info@vixonoptics.com   949-429-6363

Wangness Optics (162)
620 E 19th St #110 "Tuscon, AZ 85719"
www.wangnessoptics.com   info@wangnessoptics.com   520-661-6313
blanks/

Webster Telescopes (163)
27843 Ford Rd "Garden City, MI 48135"
www.webstertelescopes.com   info@webstertelescopes.com   734-513-2955
focusers/cells/dew shields/dob parts/drives/plans/machining services/

William Optics (164)
Taipei Taiwan
www.williamoptics.com   info@williamoptics.com   -5113
eyepieces/diagonals/prisms/adapters/filters/finders/focusers/mounts/binoviewers/

Wolf (165)
2069 Siesta Dr "Sarasota, FL 2069"
www.Ritzcamera.com   info@Ritzcamera.com
filters/eyepieces/focusers/dew shields/

Yazoo Mills (166)
PO Box 369 "New Oxford, PA 17350"
www.yazoomills.com   info@yazoomills.com   717-624-8993
tubes/

Products

Following are the products supplied by the various sources. They are listed in the form given by the suppliers. There are some overlapping categories.

abrasives
achromats
adapters
adhesives
aluminum strips-bars
artificial stars
barlows
beamsplitters
binoviewers
blanks
brass strips-bars
carry cases
cells
clutches
coating services
dew control systems
dew shields
diagonals
diagonials
dies
dob accessories
dob parts
dollies
dovetails
drives
erecting prisms
extenders
eyepieces
filters
finders
flats
flocking
flocking fabric
flocking fibers
flocking guns
flocking paper
focus motor controls
focusers
gears
glass working services
green lasers
grinding kits
guidescope rings
heat rope
heaters
knobs
large mirrors

laser collimators
lens blanks
lens cells
lens kits
lenses
machining services
mirror cells
mirror heaters
mirror kits
mirrors
mount parts
mounting hardware
mounting rings
mountings
mounts
objective lenses
observer tents
optical testing services
parabolic mirrors
piers
piggyback mounts
pitch
plans
poles
prisms
projection lenses
pyrex
quartz
raw optical glass
re-figuring services
rings
secondaries
setting circles
shroud material
shutters
special wood
spherical mirrors
spiders
t-Rings
t-adapter
taps
tools
tracking motors
tripods
tubes
used
used portholes

Supplier Codes by Product

The supplier codes refer to the alphabetical list of suppliers. This part of the listing is in the following form:

Product
xx xx xx, etc., where the xx codes refer back to the numbers in the supplier list

abrasives
61 64 100 102 105

achromats
151

adapters
6 9 16 26 33 44 53 55 57 67 74 79 82
114 115 121 132 136 137 145
152 165

adhesives
43

aluminum strips-bars
143

artificial stars
25 75

Barlows
9 12 16 23 27 32 33 35 45 48 52 53 56
63 67 68 70 72 79 82 98 111 114 116
137 140 147 152 156

beamsplitters
11 15 47 78 94 99 104 125
134 148

binoviewers
9 16 23 32 48 77 114 147 165

blanks
47 62 64 71 76 91 102 105 110 112 119
130 151 163

brass strips-bars
143

carry cases
80
cells
17 44 51 81 154 164

clutches
121

coating services
2 10 15 37 49 52 54 60 61 76 87 95 104
109 112 113 125 131 144

dew control systems
157

dew shields
17 18 25 53 65 77 82 83 153 164 166

diagonals
2 8 9 16 17 18 21 23 26 27 32 45 53 56
60 61 67 68 72 75 76 77 79 82 110 114
116 119 126 135 136 137 140 147 151
152 154 161 165

diagonials
4

dies
143

dob accessories
96

Appendix II

Telescope-Building Books and Websites

Book Must-Haves

Albert G. Ingalls, *Amateur Telescope Making Books One, Two, and Three (ATM)*, Scientific American Publications, Kingsport, TN, 1963.
(This is a collection of articles covering lots of stuff about telescope making. Most were published in *Scientific American*. Don't be fooled by the fact that this is pretty old stuff. It is good solid stuff well written and is still the best general source.)

Jean Texerau, *How to Make a Telescope, Second Edition*, Willmann-Bell, Richmond, VA, 1984.
(Takes the reader through all the steps and the whys in making an 8-inch Newtonian reflector telescope. Texereau is not an easy taskmaster. He demands that you do a good job and that you understand what you are doing. It is worth it.)

Norman Remer, *Making a Refractor Telescope*, Willmann-Bell, Richmond, VA, 2006.
(Very complete. Tells what to do and tries to be sure you know why. Includes quite a bit of optical theory to support the practice. The Ellison article in ATM is a good starting point.)

J. Warren Blaker and William M. Rosenblum, *Optics an Introduction for Students of Engineering*, Macmillan, New York, NY, 1993.
(Optical fundamentals at about the 3rd-year college level. This is not a specific telescope building book. It is a general optics book. It claims to have calculus and linear algebra as requirements, but the need is rather weak. If you don't remember that stuff you can still handle this book.)

Harold Suiter, *Star Testing Astronomical Telescopes*, Willmann-Bell, Richmond, VA, 2003.
(This is almost a companion book or extension of the Remer book.)

Davis Kriege and Richard Berry, *The Dobsonian Telescope*, Willmann-Bell, Richmond, VA, 2003.
(Detailed instructions for building large aperture telescopes.)

A.E. Conrady, *Applied Optics and Optical Design*, Dover Publications, New York, NY, 1957.
(Oldie but goodie!)

Magazines

Sky & Telescope, Sky Publishing Corp, Cambridge, MA.
Astronomy, Kalmbach Publishing Co, Waukesha, WI.

Websites

Many of these sources are, essentially, tutorials on various phases of telescope making. If you read enough of these you will have to conclude that there are a few disagreements among them. As typical of the web, some are excellent and others could use some editing for both content and language.

By the way, don't search for ATM unless you want "Automated Teller Machines." There are more of them than there are amateur telescope makers!

http://Stellafane.org/tm/atm/index.html
(Very good, many topics.)

www.Cloudynights.com
(Very good, many topics and discussion groups.)

www.webring.com/t/World-of-Astronomy-Telescope-Making/
(Lots of links.)

Natsp2000@yahoo.com
(Fiberglass tubes, focusers, mounting, large refractor)

www.scopemaking.net
(Detailed instructions for Dobsonians, mounts, observing chairs, and tripods.)

www.maryspectra.org/dobsonian/Dob.htm
(Instructions for making a large Dobsonian telescope.)

www.mdpub.com/scopeworks/index.html
(Details regarding building several types of telescopes.)

www.astropix.com/
(Once there, search on Dobsonian and 12.5 inch Dobsonian. Many other telescope items.)

www.atmsite.cog/
(Links to hundreds of useful sites.)

www.amateurtelescopemaker.com/
(ATM's resource list.)

www.starastronomy.org/telescopemaking/index.html
(Supplies and projects.)

Astro.umsystem.edu/atm/
(Large library of amateur telescope making material.)

www.astunit.com/tonkinsastro/atm/atm.htm
(UK sources and articles.)

www.aao.gov.au/local/www.sl/sl-atm.html
(Australian sources and many neat ideas.)

Theastropages.com/atm/index.htm
(Links to more links.)

Bhs.broo.k12.wv.us/homepage/alumni/dstevick/weird.html
(This is the famous weird telescope page. Don't let the name put you off; many inspired ideas.)

www.stargazing.net/mas/
(This is the web site of the Muskegon Astronomical Society, Several excellent articles on telescope building, mounts, etc.)

www.charm.net/~jriley/sky/mount1.html
(Articles of interest, including mount ideas.)

www.memphisastro.org/Mounts.html
(This is the web site of the Memphis Astronomical Society. Articles of interest.)

http://bobmay.astronomy.net/foucault/harbor1.htm
(Specific, excellent description of the meaning and use of the Foucault test. Lots of other neat stuff.)

http://www.atmsite.org/contrib/Holbrook/pvcfocuser/index.html
(Crayford focuser building.)

http://www.efn.org/~jcc/focuser.html
(Another good Crayford source.)

http://www.geocities.com/natsp2000/focuser.html
(Yet another excellent Crayford source.)

http://www.asahi-net.or.jp/~zs3t-tk/focuser/focuser.htm
(More good Crayford ideas.)

www.atmsite.org/Kohut/fgtube/index.htm/
(How to build your own fiberglass tube.)

www.woodworkingtips.com/etips/etip102000sn.html
(How to do shaper work with a router; for circle cutting.)

www.skgrimes.com/popsci/index.htm
(Describes taking lenses apart by carefully heating and re-cementing with a UV-
    activated cement.)

wapedia.mobi/en/Talk:Achromatic_lens/
(Considerable discussion of lens work.)

www.rangefinderforum.com/
(More discussion of lens restoration.)

Rfroyce.com/f8_truss_tube_newt/index.htm
(Plans, discussion of truss tubes, not Dobsonian.)

Members.cox.net/telescopemaking/index.html
(Truss tube ideas.)

www.iceinspace.com.au
(Truss tube ideas.)

www.charm.net/~jriley/sky/mount1.html
(Pipe mount plans.)

www.astro.ufl.edu/~oliver/ast3722/lectures/BasicScopes/BasicScopes.html
(Course notes; includes good stuff on mount design.)

Appendix III

Polishing Log

This section will provide an actual log of hand polishing an optical surface from fine grind to a reasonably good surface. The surface described is the back side of the flint component of the lens for the 4-inch f17 instrument described in Chap. 7.

This material is presented because you might be re-polishing because of an excess of pits in a used lens or mirror. By examining the surface of your glass and the comments column of this log you may be able to fit the condition of the surface to some stage of polishing from a fine grind surface. Or, you may be re-polishing as part of a re-figuring process for a mirror with a bad figure. The same observation applies. Whatever the source, the process of producing an optical quality polish can take quite a bit of time and effort. Note that the example used just 1 min less than 11 h.

The exact specifications for the complete lens are as follows:

- The crown component has an index of refraction of 1.517 and dispersion of 64.5. This is normally written as a fraction, i.e., 1.517/64.5.
- The flint component would be designated as 1.62/36.2.
- The crown radii are R1 = R2 = 31.303, producing a focal length (just for that component) of 30.27 inches. Note that the crown and flint were ground on each other. The radius difference between the two occurs during polishing of the crown.
- The flint radii are R3 = −31.503 R4 = +489.15
- The polishing log is for the R4 side with a very long radius. Polishing on pitch lap with rouge.

Start time	End time	Time (in min)	Total time	Remarks
5:00	5:15	15	15	Sticking, getting little flat spots between pits
6:15	6:45	30	45	Works OK now Cold press all night
10:45	10:47	2	47	Handle came off
11:01	11:31	30	77	
3:40	4:10	30	107	Cold press all night
12:00	12:30	30	137	
3:44	4:14	30	167	
5:37	5:52	15	182	Much better in the center. Still very pitty at edges. Conclusion: At least not making a turned edge
7:08	7:23	15	197	Turn over to lap on top
1:30	2:00	30	227	
4:53	5:00	7	234	
5:17	5:52	35	269	Edges are better but still milky. Another 1 1/2 h should get it looking superficially good and three more actually good
7:30	8:00	30	299	
9:53	10:23	30	329	
11:55	12:25	30	359	Still a lot of pits at the edges. Much better than 1 1/2 h ago
2:23	2:53	30	389	
5:02	5:32	30	419	
10:41	11:21	30	449	Superficially looks good. Strong magnifying glass still shows quite a few pits in the outer 1/4 inch. Need three more hours on this surface?
10:16	10:46	30	479	
6:36	7:06	30	509	
11:04	11:34	30	539	Assemble the lens and try on stars. No moon On Arcturus I get nice clear diffraction rings on both sides of focus. So some more polishing is appropriate Easily split Mizar sep = 14". Could easily get 1/4 of that Tried the double double 2.35". No problem! Pressed 1 h
10:20	10:40	30	569	
2:15	2:45	30	599	
7:24	7:54	30	629	
10:35	11:05	30	659	Still a few pits at extreme edge. But it will be behind the ring and covered

Appendix IV

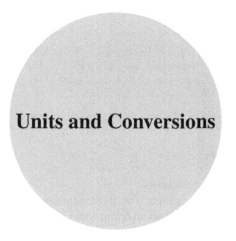

Units and Conversions

General Conversions

1 centimeter = 0.3937 inches
1 inch = 2.54 centimeters = 25.4 millimeters
1 pound = 0.4536 kilograms
1 kilogram = 2.2046 pounds

Wood Stock and Hardware Conversions

European Plywood Standards

4 mm	a little less than 1/4 of an inch; usable replacement for 1/4
9 mm	a little less than 1/3 of an inch; usable replacement for 3/8
10 mm	almost 4/10 of an inch; also usable replacement for 3/8
12 mm	almost 1/2 of an inch; usable replacement for 1/2
15 mm	almost 5/8 of an inch; usable replacement for 5/8
18 mm	almost 3/4 of an inch; usable replacement for 3/4

Timber

45 × 70 mm	quite a bit less than a 2 by 4; usable replacement for 2 by 4
42 × 95 mm	not so much less than a 2 by 4; even better replacement
60 × 100 mm	more than a 2 by 4; oversize replacement for a 2 by 4

Hardware

Sheet metal sold by metric thickness.

Machine screws sold by gauge. The most common one used in this book is 1/4 × 20, which works out to be about 6 mm coarse thread.

Index

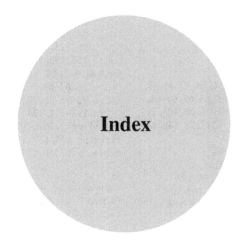